The Cybersecurity Fear Machine

The Cybersecurity Fear Machine

How Inflated Threats Hijacked America's Critical Infrastructure Policy

Daniel Ward, Ph.D.

Fountain & Key Press · Cambridge, MA

The Cybersecurity Fear Machine: How Inflated Threats Hijacked America's Critical Infrastructure Policy

ISBN (Paperback): 978-1-972842-01-0

ISBN (Ebook): 9781972842003

First Edition

Published by Fountain & Key Press

Cambridge, Massachusetts

fountainandkeypress.com

Artificial intelligence tools may have been used to support the writing, editing, research, and/or production of this work. All content has been reviewed and approved by the author, who is responsible for the final text.

Printed in the United States of America

Contents

Chapter 1

THE UNEXAMINED NARRATIVE: OPERATIONAL TECHNOLOGY UNDER SIEGE?

On May 7, 2021, an employee at Colonial Pipeline discovered a ransom note on a company computer. The company, which by its own account supplies roughly 45 percent of the fuel consumed on the U.S. East Coast, was hit by a ransomware attack. The attackers had not breached the operational technology (OT) controlling the pipeline itself, but had compromised the company's information technology (IT) billing systems. Fearing the infection could spread and unable to bill customers, Colonial's leadership made the extraordinary decision to shut down the entire 5,500-mile pipeline — the first full system shutdown in the company's history attributable to a cyber incident.

The effects were immediate. Panic buying led to widespread gasoline shortages, flight cancellations, and a spike in fuel prices. The roughly five- to six-day shutdown was not caused by a sophisticated cyber-physical attack, but by an IT breach that prompted an operational shutdown out of caution. The incident illustrates the book's central theme: while the discourse on OT security focuses on catastrophic physical destruction, the most significant impacts often stem from failures in governance, resilience, and the blurred lines between IT and OT.

A potent narrative dominates cybersecurity policy: America's critical infrastructure is on the precipice of catastrophic cyber-physical attack. This book argues that the discourse is misaligned with the evidence. It overemphasizes catastrophic attacks while underappreciating more frequent threats like governance failures and poor resilience, directing resources toward improbable scenarios instead of foundational vulnerabilities.

To deconstruct this narrative, we must first define "Operational Technology" (OT). Unlike Information Technology (IT), which focuses on data processing, communication, and business operations, OT systems are designed to monitor and control physical processes. Think of the Supervisory Control and Data Acquisition (SCADA) systems that manage power grids, the Distributed Control Systems (DCS) that operate chemical plants, or the Programmable Logic Controllers (PLCs) that automate manufacturing lines. These are the digital interfaces to the physical world, where failures have direct, tangible consequences. A compromised power grid can lead to widespread blackouts, a disrupted water treatment facility can endanger public health, and a halted manufacturing plant can disrupt supply chains. The security of OT is a distinct challenge due to its real-time operational needs, long system lifecycles, and the direct physical impact of failure.

The Genesis of the Narrative

The narrative of imminent cyber-physical catastrophe did not emerge in a vacuum; it evolved from technological advancements, geopolitical events, media portrayals, and the rise of a powerful cybersecurity industry. Early, often sensationalized, cyber incidents etched a sense of vulnerability into the public consciousness. The initial reporting of computer viruses and network intrusions, while often lacking detail about the actual impact on physical systems, planted the seed of digital danger. As the internet became more pervasive and interconnected, the potential attack surface expanded, and so did the perceived threat. Geopolitical tensions further amplified this narrative, with accusations of state-sponsored cyber espionage and sabotage becoming commonplace, even when attribution was uncertain or politically motivated. This historical trajectory, which we will explore in more detail in subsequent sections, laid the fertile ground for the "fear machine" to take root.

What is striking about this narrative is its age. The prediction of civilizational collapse by keystroke is not a product of the ransomware era; it is older than the commercial internet itself. The specter of an "Electronic Pearl Harbor" was invoked as early as 1991, long before most Americans had ever sent an email.[1] Two decades on, the warning had migrated from think-tank papers to the highest levels of government. On October 11, 2012, in a speech to Business Executives for National Security aboard the USS Intrepid, Secretary of Defense Leon Panetta warned that a 'cyber Pearl Harbor' could derail passenger trains, or even more dangerous, derail trains loaded with lethal chemicals, and could 'contaminate the water supply in major cities, or shut down the power grid across large parts of the country.'[2] These were not abstractions; they were specific, vivid, catastrophic scenarios attached to a deadline that was always imminent. And yet the central, uncomfortable fact this book sets out to examine is that more than three decades after the phrase was coined, and well over a decade after Panetta's warning, there has been no cyber-induced derailment of a chemical train, no cyber-caused mass poisoning of a city's water supply, and no U.S. grid felled by an adversary's keystroke at civilizational scale — though intrusions or suspected intrusions into water utilities (the disputed Oldsmar, Florida incident in 2021; the CyberAv3ngers PLC compromise at Aliquippa, Pennsylvania in 2023) show the surface is real even where consequences have been contained, and Russian-attributed cyberattacks caused localized grid outages in Ukraine on December 23, 2015 (affecting regional distribution utilities) and on the night of December 17–18, 2016 (the Pivnichna transmission substation north of Kyiv), demonstrating that limited, targeted disruptions are possible even if the catastrophic civilizational version has not materialized. We will return to this gap between prophecy and record again and again, because it is the thread from which the entire fear machine is woven.

Understanding the "fear machine" requires us to identify its architects and their motivations. It is not a monolithic entity but an ecosystem of actors who amplify perceived threats. Several groups contribute to this ecosystem:

- **Cybersecurity Vendors:** Have a vested interest in highlighting threats to sell products and services.
- **Consultants:** Often emphasize worst-case scenarios to justify risk management engagements.
- **Think Tanks:** Can shape policy discourse with reports, sometimes funded by industry or government, that create demand for specific security measures.

- **Government Agencies:** May use perceived threats to justify budgets, expand mandates, and demonstrate their role in national security. Recognizing these diverse motivations is crucial for dissecting the "fear machine" and understanding how commercial interests and bureaucratic imperatives can shape public perception and policy decisions, potentially diverting resources from more pressing, fundamental issues.

The scale of the commercial interest at stake is not incidental to this story; it is the engine of it. Industry analysts project the global cybersecurity market will reach roughly $322 billion by 2029.[3] A market of that magnitude does not merely respond to fear; it has every reason to cultivate it. We will hold that number in mind throughout what follows, because the question of who profits from the narrative of perpetual siege is one no honest accounting of OT security can avoid.

Several characteristics distinguish OT from IT and contribute to its security challenges. Firstly, OT systems often have extraordinarily long lifecycles, frequently measured in decades, far exceeding the three- to five-year refresh cycles typical for much enterprise IT hardware. This longevity means many OT systems are built on outdated technologies, operating systems, and communication protocols that were designed with little or no consideration for cybersecurity. Patching and upgrading these systems are often complex, costly, and disruptive, requiring extensive downtime or specialized engineering expertise that may not be readily available. Secondly, OT systems are characterized by stringent real-time operational requirements. Unlike IT systems where a slight delay in data processing might be acceptable, in OT environments, even milliseconds of latency can have severe consequences, potentially leading to equipment damage or unsafe operating conditions.

This makes traditional IT security measures, such as frequent reboots or intrusive scanning, impractical or even dangerous. Finally, the direct link to physical processes means that cybersecurity failures in OT can result in immediate, tangible, and potentially catastrophic physical damage, injury, or environmental harm. This differs from most IT breaches, which typically involve data theft or business disruption. It is this direct physical impact that fuels much of the fear surrounding OT cybersecurity, and it is precisely this aspect that has been heavily leveraged in the construction of the "fear machine."

Government agencies also participate. Justifying large budgets and expanded mandates often relies on demonstrating a significant threat. Cybersecurity has become a vital component of national security, and highlighting potential vulnerabilities allows agencies to secure resources and assert their importance in protecting the nation. This can create a feedback loop: perceived threats are amplified to secure funding, which fuels demand for security products and reinforces the narrative. Understanding these motivations, the commercial imperative of vendors, the economic incentives for consultants, the influence of think tanks, and the bureaucratic needs of government, is essential for recognizing how the "fear machine" operates and how it can shape policy and public perception in ways that may not always align with a pragmatic assessment of actual risks. The real vulnerabilities, as we will explore, often lie in less sensational, but more fundamental, areas of governance and operational practice.

Our analytical framework for deconstructing the "fear machine" is grounded in a commitment to empirical data, rigorous historical analysis, and a critical examination of policy and industry claims. We reject speculative threat assessments and sensationalized narratives in favor of evidence-based reasoning. Our methodology relies on several key pillars. Firstly, we undertake a forensic examination of documented cyber incidents targeting U.S. OT infrastructure. This involves scrutinizing the scope, impact, and attribution of notable events to establish a factual baseline of actual damage, rather than relying on hypothetical scenarios or vendor-driven threat intelligence. We will assess the frequency and severity of past attacks to challenge the notion that catastrophic, nation-state-level attacks are common. Specific, verifiable incidents will be detailed, analyzing their true consequences and the lessons learned, or perhaps, not learned.

Secondly, we draw a crucial distinction between simulated exercises, vendor demonstrations of potential capabilities, and actual, successful breaches that resulted in significant disruption or damage. It is a common tactic within the "fear machine" to conflate theoretical vulnerabilities or exercises with historical realities. We will highlight how "near-misses" or the potential for exploitation, while important for awareness, are frequently presented as de facto successes or imminent threats, thereby inflating perceived risk. This analysis aims to clarify what constitutes a genuine cybersecurity failure versus a demonstration of potential risk, underscoring the importance of empirical evidence over hypothetical scenarios.

The interaction between these actors, commercial vendors, consultants, think tanks, and government agencies, creates a complex and often self-reinforcing ecosystem. Each has a vested interest, whether financial, political, or institutional, in maintaining and, at times, amplifying the perception of an escalating OT cybersecurity threat. This does not negate the reality of OT threats themselves, which are significant and evolving; what we contest is the catastrophic framing, not the existence of risk. However, understanding these underlying motivations is crucial for dissecting the "fear machine" that has become so deeply embedded in the public and policy discourse surrounding operational technology. It allows for a more critical evaluation of the information presented, enabling stakeholders to distinguish between genuine security needs and the commercial or political imperatives that may shape the articulation of those needs. By recognizing who benefits from the narrative of constant siege, we can move towards a more balanced, evidence-based approach to securing OT systems, one that prioritizes effective risk management over uncritical acceptance of amplified anxieties.

An Evidence-Based Approach

This book's analysis is grounded in empirical evidence. The methodology rests on three pillars:

- **Forensic Examination of Incidents:** We analyze documented OT incidents to establish a factual baseline, distinguishing verifiable events from hypothetical scenarios.
- **Distinguishing Capability from Impact:** We distinguish demonstrated capabilities (e.g., proofs-of-concept) from successful attacks causing disruption, as the two are often conflated.
- **Root Cause Analysis:** We investigate governance failures, human error, and poor maintenance as frequent causes of OT disruption, comparing their impact to malicious cyber intrusions.

This thesis would be challenged by clear evidence of widespread, catastrophic OT incidents caused by external adversaries.

What to Expect

The book is organized into four parts. Part One establishes an empirical baseline for OT incidents and governance. Part Two dissects the "fear machine" and its incentives. Part Three tests the catastrophe narrative against evidence. Part Four proposes reforms for a more resilient, evidence-based security posture.

CHAPTER 2

THE HISTORICAL RECORD: SCARCITY OF CATASTROPHE

The historical record of cyberattacks on Operational Technology (OT) is not one of constant catastrophe. While the potential for severe disruption is real, documented events reveal targeted intrusions and disruptive cybercrime rather than catastrophic outcomes. This chapter scrutinizes notable incidents to establish a factual baseline.

But establishing that baseline requires more than counting incidents. It requires reading each one twice: once as it is *commonly told*, and once as the *documented record* actually supports. The gap between those two readings is the subject of this chapter. Almost every famous OT cyber incident arrives in public discourse pre-loaded with a dramatic shape, a villain, a near-apocalypse averted at the last second. And almost every one, examined against primary sources, resolves into something narrower, stranger, and more instructive than the headline. The pattern is not that these events were fabricated. The pattern is that they were *mislabeled*, with the load-bearing details quietly inverted. A criminal extortion becomes a national security strike. An IT outage becomes a physical hack. A single contested phone call becomes a permanent fact. A laboratory test becomes documentary footage of a real attack. Where the previous chapter framed the problem in the abstract, this chapter does the forensic work case by case, distinguishing the narrative we inherited from the evidence we can actually point to.

A Framework for Analysis

This chapter uses a consistent framework to assess the historical record: a severity rubric, a table of documented incidents, and an acknowledgment of analytical limits. For each headline case, it also applies a two-column discipline, *narrative as commonly told* versus *what the documented record shows*, so that the reader can see exactly where the popular account departs from the sourcing. The goal is not to declare that nothing happened. It is to insist that what happened be described in the words the evidence will bear, and no stronger.

A Note on Methods and Limits Analyzing OT cyber incidents presents several challenges: **Underreporting:** Many incidents are never publicly disclosed due to reputational risk, legal concerns, or a lack of reporting requirements. The available data is an incomplete subset of all events. **Attribution Uncertainty:** Pinpointing the responsible actor is difficult, as evidence can be obscured and attribution often relies on classified intelligence or geopolitical inference rather than definitive technical proof. **Classification:** The line between an IT and OT incident can be blurry, especially when an IT breach (like ransomware) forces an OT shutdown as a precautionary measure.

The incidents in this chapter are drawn from primary sources (company disclosures, government alerts) and reputable secondary sources (cybersecurity firm reports, investigative journalism). While every effort is made to rely on verifiable data, some details regarding impact or attribution remain subject to the limitations outlined above.

One further caveat cuts the other way, and intellectual honesty requires stating it plainly. Some respected practitioners, the control systems engineer Joe Weiss prominent among them, argue that OT physical-impact incidents are *under-counted* rather than over-counted, precisely because OT environments often lack the forensic logging that would let an investigator distinguish a cyber cause from an equipment fault.[1] This is a fair methodological point and it is taken seriously here. The response is not to dismiss it but to be precise about scope: most of the incidents Weiss and others classify as under-counted are equipment-level malfunctions or insider mistakes, not the scenario the public fears, a remote adversary reaching across a network to seize physical controls and cause harm. Both things can be true at once. The long tail of murky, unlogged process upsets may indeed be larger than the public record suggests, *and* the specific, headline-grade, remote-attacker-destroys-infrastructure event remains conspicuously rare in the documented record. This chapter concerns the second claim.

OT/ICS Incident Severity Rubric

Level	Category	Description	Example
1	Nuisance	Minor operational anomalies or data exposure with no disruption to physical processes or safety.	Unauthorized network scan; temporary loss of view on an HMI.
2	Disruption	Localized and temporary disruption of a physical process; contained financial loss.	A single production line halted for several hours; minor data breach.
3	Significant Disruption	Widespread or prolonged disruption to a key service or production; significant financial loss; localized public impact.	Colonial Pipeline (2021): IT ransomware attack forces multi-day precautionary shutdown of OT.
4	Severe Disruption	Major, prolonged outage of critical infrastructure with regional impact; risk of injury or localized environmental damage.	Ukraine Grid Attacks (2015/2016): Coordinated attacks cause power outages for over 200,000 people for hours.
5	Catastrophic	Widespread, long-term failure of critical infrastructure resulting in mass casualties, severe economic collapse, or a national security crisis.	*No documented historical examples directly caused by a cyberattack.* Classification requires verifiable public evidence from national authorities, international bodies, or corroborated reporting from multiple credible journalistic sources. Speculation does not meet this threshold.

Date	Sector	Type	Impact	Downtime	Source(s)
2010	Nuclear	OT-Native (Sabotage)	Physical destruction of ~1,000 centrifuges.	N/A (Degradation)	Symantec, Langner
2014	Manufacturing (Germany)	IT→OT	Physical damage to a steel mill blast furnace.	Unspecified	BSI (Germany)
2015	Electric Utility (Ukraine)	OT-Native (Disruption)	Power outage for ~225,000 customers.	~6 hours (manual recovery)	E-ISAC, SANS
2016	Electric Utility (Ukraine)	OT-Native (Disruption)	Power outage in Kyiv.	~1 hour	Dragos, ESET

Date	Sector	Type	Impact	Downtime	Source(s)
2017	Petrochemical (Saudi Arabia)	OT-Native (Sabotage)	Targeted SIS malware that could have enabled unsafe conditions; discovered after a fail-safe shutdown; no explosion, casualties, or public physical harm.	None (tripped to safe)	FireEye, Schneider Electric
2021	Water & Wastewater (USA)	IT→OT	Attempted manipulation of water chemical levels; later unconfirmed; no public impact.	None	Pinellas County Sheriff; FBI (later unconfirmed)
2021	Pipeline (USA)	IT→OT (Ransomware)	Precautionary shutdown of major fuel pipeline; public evidence does not show OT compromise; panic-buying drove shortages.	~5 days	Colonial Pipeline, FBI, DOJ
2023	Water & Wastewater (USA)	IT→OT	Defacement of internet-exposed PLCs (default password); no public-health impact and no community loss of drinking-water service, though one booster station shifted to manual operation.	Minimal	CISA, Multiple Utilities
2021	Food & Agriculture (USA)	IT→OT (Ransomware)	Shutdown of meat processing plants, disrupting food supply chain.	~1 day	JBS, FBI
2022	Manufacturing (Japan/Global)	IT→OT (Ransomware)	Toyota suspended all 14 domestic plants (28 production lines) after supplier Kojima Press Industry was hit by ransomware.	~1 day	Reuters, Feb 28, 2022; Toyota corporate statement

The table above lists the events most often cited as proof that the sky is falling. What follows is the careful reading of each, the part the citations usually skip. Read in this two-column way, the canonical set of "infrastructure cyberattacks" tells a consistent story: real intrusions, real expense, real lessons, and almost no instance of a remote adversary actually wresting physical control of a critical process and causing public harm.

Stuxnet (2010): the superpower's scalpel, sold as everyone's knife. The Stuxnet worm, discovered in 2010, brought OT cybersecurity into stark relief, and it remains among the most technically accomplished cyber-physical operations ever made public. *As commonly told*, Stuxnet is the proof that malicious code can reach out and physically wreck industrial equipment, an open-ended warning that what happened in Iran could happen to any plant with a programmable logic controller. *What the documented record shows* is narrower and, for the purpose of assessing commodity threats, almost the opposite of the popular lesson. Stuxnet was a nation-state operation, widely reported as a joint U.S.–Israeli program codenamed "Olympic Games," developed over years at a cost that only a government could absorb.[2] It chained together multiple Windows zero-day vulnerabilities, used stolen, legitimate digital certificates to sign its drivers and appear trustworthy, and was engineered to jump an air gap into a network with no direct internet connection, reportedly via removable media. Its payload was not generic. It searched specifically for Siemens S7 controllers driving a particular configuration of centrifuge cascade, and ignored everything else. Its target was singular: the bunkered, air-gapped enrichment hall at Natanz. There it damaged on the order of 1,000 of the roughly 9,000 IR-1 centrifuges then installed at Natanz while replaying recorded, falsified "normal" sensor readings back to operators, so the machines tore themselves apart while the control room saw nothing wrong.[3] Read correctly, Stuxnet is evidence of what a superpower can do to *one* fixed, high-value target, with enormous effort, bespoke intelligence, and years of preparation. That is a poor template for the commodity threats that ordinary utilities and manufacturers actually face. The barriers to entry that Stuxnet revealed, the zero days, the stolen certificates, the physical intelligence about a specific cascade, the air-gap-jumping logistics, are precisely why it was *not* the opening shot of a global sabotage wave. In the fifteen years since, no cyber-physical weapon of comparable sophistication and destructive effect has been documented against civilian infrastructure. The worm is rightly a landmark; its abuse comes from treating a state's scalpel as if it were a knife any criminal could pick up.

Ukraine grid (2015 and 2016): the famous "blackout hacks" that lasted hours because operators drove to the substations. The Ukrainian power grid attacks of 2015 and 2016 are the first widely confirmed cyberattacks to cause electrical outages, and remain the canonical examples, and they are state-sponsored work, attributed to the Russian military intelligence group commonly tracked as Sandworm, carried out against a country with which Russia was already at war.[4] *As commonly told*, they are the proof that a grid can simply be switched off by remote attackers, the dress rehearsal for an

American blackout. *What the documented record shows* is a far more sobering lesson for anyone selling grid-collapse scenarios. In December 2015, attackers used stolen credentials to operate the distribution utilities' own control software, opening breakers and cutting power to roughly 225,000 customers; to slow the response, they also overwrote firmware on serial-to-Ethernet devices and flooded the call center with automated phone calls so customers could not report the outage.[5] And yet customer power was restored within roughly one to six hours (though full automated operation was not restored for months). Not because the defenders out-hacked the attackers in real time, but because operators fell back to *manual* control: they physically drove to the affected substations and closed the breakers by hand.[6] The 2016 follow-on, using the purpose-built ICS malware variously called Industroyer or CrashOverride, was more automated and more ambitious in design, capable of speaking grid protocols directly, yet in practice it knocked out a single transmission substation north of Kyiv for roughly an hour and a quarter.[7] The headline writes itself: the most famous grid hacks in history, executed by a nation-state's military hackers, in the middle of a shooting war, kept the lights off for *hours*, because the people running the system could still throw the switches themselves. That manual fallback, the very "backwardness" that grid-modernization marketing often treats as a liability, was the resilience that bounded the damage.

TRITON / TRISIS (2017): "the world's most murderous malware," discovered because it failed safe. In 2017, malware later named TRITON (also TRISIS) was found inside a petrochemical facility in Saudi Arabia, where it targeted Schneider Electric Triconex Safety Instrumented Systems, the dedicated controllers whose only job is to bring a plant to a safe state if a process runs out of bounds. Because tampering with a *safety* system is the closest any documented attack has come to courting human casualties, TRITON has been described in the press as "the world's most murderous malware." The attribution points to a nation-state: the threat intelligence firm FireEye (now Mandiant) assessed that the activity was supported by a Russian government research institute, the Central Scientific Research Institute of Chemistry and Mechanics (CNIIHM), and in October 2020 the U.S. Treasury sanctioned CNIIHM in connection with the attack.[8] *As commonly told*, TRITON is the moment the nightmare nearly arrived, an attacker reaching for the kill switch on a hazardous plant. *What the documented record shows* is that the operation was discovered only because it *failed*, and failed *safe*. The intruders' code inadvertently tripped the Triconex controllers into a protective shutdown, halting the plant. That shutdown, an engineering nuisance, was what exposed the entire campaign.[9] The "disaster," in other words, was the safety system doing exactly what it was built to do. The lesson is not that safeguards are futile; it is that even a state actor sophisticated enough to reach a safety controller was defeated by the layered, fail-closed design philosophy that governs how hazardous facilities are engineered.

Colonial Pipeline (2021): the pipeline that shut itself off because the billing system was down. In the United States, major OT cyberattacks causing catastrophic damage have been notably absent. Instead, the historical record shows a prevalence of less sensational, though still significant, types of intrusion, and no case illustrates the gap between narrative and record more cleanly than the May 2021 Colonial Pipeline incident. *As commonly told*, hackers, often imagined as a hostile government, seized control of the largest fuel pipeline on the U.S. East Coast and shut it down, triggering fuel shortages from a direct strike on critical infrastructure. *What the documented record shows* dismantles nearly every load-bearing element of that account. The attackers were the DarkSide group, a financially motivated ransomware-as-a-service criminal operation, not a nation-state; President Biden stated publicly that there was "no evidence" the Russian government was responsible.[10] The intrusion did not breach the pipeline's operational technology at all. It encrypted Colonial's *IT* and billing systems. The pipeline's control systems were never hacked, Colonial shut the line off *itself*, preemptively, in substantial part because the ransomware had locked up the billing system and the company could not invoice for the fuel it moved.[11] The single point of failure was mundane: entry came through one legacy VPN account, protected by a password and no multi-factor authentication.[12] The fuel shortages that dominated the coverage were driven substantially by *panic-buying*, drivers topping off tanks and filling jerricans, rather than by any physical scarcity the shutdown alone had yet created. And the ending rarely makes the retelling: the Department of Justice subsequently clawed back 63.7 of the roughly 75 bitcoin Colonial had paid, recovering most of the ransom in bitcoin terms (though a smaller share of the dollar value originally paid, owing to the bitcoin price decline).[13] The sharpest possible summary is this: *the pipeline's control systems were never hacked, Colonial shut it off itself, largely because ransomware had locked up the billing system.* This was a serious, costly, genuinely disruptive crime. It was not a cyber-physical attack on a pipeline, and the distinction is the whole point. Critical infrastructure can suffer severe consequences from the *interconnectedness* of IT and OT and from the business decisions that interconnection forces, even when the OT itself is never touched. The event rightly prompted new federal directives for pipeline operators and a renewed focus on segmenting IT from OT, on backups, and on the unglamorous matter of turning on multi-factor authentication.

Oldsmar, Florida (2021): the water-poisoning that senior officials later called a "non-event." The water and wastewater sector, despite being critical infrastructure, has also generated headline scares that shrink under scrutiny, and the 2021 incident at the water treatment plant in Oldsmar, Florida is the cautionary tale about how a single dramatic account hardens into permanent "fact." Here the reality-check framing must be handled with particular care, because the honest position is not that we know what happened, but that an external attack was never established and the original account was later walked back by the officials closest to it.

As commonly told, an unknown intruder remotely accessed the plant's controls and tried to poison the town's water by raising the level of sodium hydroxide, lye, to dangerous concentrations; an alert operator, watching his screen, saw the cursor move on its own and dragged the setting back down just in time. That vivid story traces to essentially *one source on one day*: a press conference by Pinellas County Sheriff Bob Gualtieri shortly after the event.[14] *What the documented record shows* in the years since is a striking retreat from that narrative. Roughly two years later, Oldsmar's city manager, Al Braithwaite, publicly characterized the episode as a "nonevent," saying the FBI had found "no evidence of any access from the outside" and that it was "likely the same employee" who had been "actually banging on his keyboard," rather than an intruder.[15] CyberScoop, reviewing the matter, reported that "the FBI was not able to confirm that this incident was initiated by a targeted cyber intrusion".[16]

Two points must be stated precisely, in both directions. First, this book does *not* assert that Oldsmar was definitively operator error. The FBI's conclusions were never publicly released in full, and certainty is not available; what can be said is that an external intrusion was never established, that senior officials later walked the dramatic version back, and that the matter remains, at best, unproven. Second, even taking the original, most alarming telling entirely at face value, no public harm was ever remotely close. By the sheriff's own account, the altered sodium-hydroxide setpoint "would have taken from 24 to 36 hours to take effect," and the change "would have triggered" the plant's pH alarms long before any tainted water reached a customer.[17] No water was contaminated. No one was endangered. The engineered safeguards, time delay, pH alarming, downstream monitoring, made public harm effectively impossible regardless of who moved the setting or why. Oldsmar thus does double duty as a cautionary tale: it warns about exposed remote-access on control systems, and it warns, just as loudly, about how a single uncorroborated account can become an unkillable data point in the threat narrative.

Cyber Av3ngers and Unitronics (2023–24): defacing PLCs with the password "1111." A more recent water-sector scare deserves the same treatment because it was widely reported as foreign hackers attacking American water systems. In late 2023 and into 2024, a group calling itself Cyber Av3ngers compromised Unitronics programmable logic controllers at water and other facilities. *What the documented record shows* is that the affected devices were internet-exposed and were reachable using the Unitronics default password, "1111"; the intruders changed the human-machine interface screens to display a political message.[18] It was closer to defacement than to sabotage: beyond changing the HMI screens, the intruders disrupted a booster station's pressure-regulating logic, forcing a switch to manual operation — a real but contained OT impact, with no documented loss of water service to any community. The episode is a real indictment of basic hygiene, devices that should never face the open internet, default credentials left unchanged, but it is not the sophisticated assault on the water supply that the headlines implied.

The German steel mill (2014): the canonical "physical destruction" case is a single anonymous paragraph. If any incident is treated as the bedrock proof that a cyberattack can physically destroy heavy industrial equipment, it is the German steel mill. *As commonly told*, hackers breached a steel plant, took control of the systems, and caused a blast furnace to be shut down improperly, inflicting "massive" physical damage, the closest thing on the books to a deliberate cyber-physical wrecking of a factory. *What the documented record shows* is far thinner than that reputation, and the gap here is the most important in the chapter. The core public evidence for the steel mill case still rests primarily on BSI's sparse account: a *single anonymous, undated, untechnical paragraph*, roughly page 31, of the German Federal Office for Information Security's (BSI) 2014 annual report.[19] That paragraph names no company. It includes no malware sample, no forensic report, no indicators of compromise, no quantified figure for the damage, and no independent verification. In the eleven-plus years since, no public forensic report, malware sample, indicators of compromise, or quantified damage figure has ever publicly surfaced; later secondary analyses, including DOE/INL's CyOTE work, have inferred a likely victim and timeline but have not produced a full technical writeup.

The intellectually honest conclusion is *not* that the event did not happen. It may well have. The conclusion is about the *thinness of the evidence*: the canonical proof that hackers can physically destroy industrial equipment is, to this day, a single anonymous paragraph in a government report, unaccompanied to this day by any public primary forensic report, malware sample, or quantified-damage figure. A claim that load-bearing deserves load-bearing evidence, and this one has never had it. That an entire genre of "they can melt your factory" argumentation rests on one unsourced paragraph is itself a finding about the state of the evidence, and it is exactly the kind of soft spot that the broader threat narrative is built to keep readers from noticing.

Aurora (2007): a laboratory test, replayed for years as if it were an attack. A final, instructive entry is not an incident at all. The Aurora Generator Test was a 2007 experiment conducted by the Idaho National Laboratory, in which researchers used a cyber technique to open and reclose a diesel generator's breakers out of phase, destroying the machine in a cloud of smoke and metal under controlled conditions.[20] The footage of that generator shuddering itself to death is genuinely arresting, which is precisely why it has been recycled for years in presentations and broadcasts, sometimes presented, or simply allowed to be understood, as if it were documentation of a real attack on the grid. It was a *test*, a demonstration of a real vulnerability class staged by defenders to study it, not an adversary striking a live target. Treating lab footage as field evidence is the demonstration-versus-incident confusion in its purest visual form.

Volt Typhoon (2023–24): real, current, state-backed, and not the same thing as an attack. The most contemporary entry must be handled with the same precision, in the opposite direction from the cases above, because here the danger is *under*-reacting to something real. Beginning in 2023 and continuing into 2024, U.S. authorities warned that People's Republic of China state actors, tracked as Volt Typhoon, were "pre-positioning" themselves on the IT networks of U.S. critical infrastructure organizations, maintaining quiet, persistent access apparently to enable disruption at a future moment of Beijing's choosing, per the joint CISA advisory AA24-038A.[21] This is not fake, not overstated, and not a vendor demonstration. It is a genuine, ongoing, nation-state campaign and it belongs in any sober threat assessment. The discipline this chapter demands is simply to describe it in the words the evidence supports, and the operative distinction is *access versus impact*. Pre-positioning and persistent access are not the same as disruption. As of this writing, no physical effects, no outages, and no destructive attacks have been attributed to Volt Typhoon. The campaign is correctly understood as alarming *preparation*, the establishment of a foothold, and it warrants serious defensive response on exactly those terms. Reporting it as a blackout that has already happened would repeat the very error this chapter is cataloguing; ignoring it because no blackout has happened yet would be the mirror-image mistake.

Two further categories round out the documented record without changing its shape. The first is the theft of intellectual property and operational data from OT-adjacent environments, frequently attributed to state-sponsored espionage, intrusions aimed at *learning*, not breaking. Reports have recurred for years of state-linked groups targeting U.S. energy companies, chemical manufacturers, and other industrial entities to exfiltrate proprietary information over long periods.[22] The long-term costs to economic competitiveness and national security are real and hard to quantify, but the activity is intelligence-gathering, not the immediate, widespread physical destruction the word "catastrophe" conjures. The second is the cluster of *ransomware-driven IT→OT shutdowns* that fill out Table 2.1, JBS (food processing, ~1 day) and Toyota's Japanese plants halted after a ransomware hit on its supplier Kojima Industries (~1 day).[23] Each was disruptive and costly. In each, as at Colonial, the harm flowed from a business decision to halt operations after an *IT* compromise, not from an adversary manipulating physical controls.

Why is catastrophe so rare in the documented record? Several factors contribute:

- **System Design:** Many OT systems remain segmented from IT networks or use specialized protocols, making remote access difficult. They are also built with physical safety systems, redundancies, and manual overrides designed to prevent catastrophic failure. The Ukrainian operators who drove to their substations, and the Triconex controllers that tripped TRITON's own plant into a safe state, are not anecdotes; they are the mechanism by which severe attacks were bounded into hours-long or no-harm outcomes.

- **Attacker Requirements:** A successful cyber-physical attack requires immense resources, expertise, and specific intelligence about the target. This sophistication is typically limited to a few nation-state actors, who risk attribution and retaliation. Stuxnet is the proof of the price of admission: years of work, multiple zero-days, stolen certificates, and bespoke knowledge of one cascade, spent against a single fixed target.

- **Adversary Motives:** The most common cyber threats are driven by financial gain (ransomware) or intelligence gathering (espionage), not physical destruction. These motives lead to disruptive or costly incidents, but not necessarily catastrophic ones. The criminals behind Colonial and JBS wanted to be *paid*; breaking a pipeline or a slaughterhouse would have destroyed the leverage they were selling back.

- **Defensive Efforts:** While imperfect, cybersecurity defenses are continuously improving. Increased awareness, investment in security, and government regulation have hardened many potential targets, though, as Oldsmar's exposed remote access and Cyber Av3ngers' "1111" passwords show, the gap between what we know to do and what is actually done remains wide.

In conclusion, the historical record reveals a landscape of persistent threats, disruptive incidents, and significant espionage, but a notable absence of catastrophic cyber-physical attacks on U.S. critical infrastructure. Read case by case, the famous exceptions prove the rule: the superpower's scalpel at Natanz, the wartime grid hacks undone by hand, the "murderous" malware caught failing safe, the pipeline that shut itself off, the water-poisoning that senior officials later called a nonevent, and the "physical destruction" case that survives only as one anonymous paragraph. This evidence counters narratives that overemphasize doomsday scenarios and supports security strategies focused on resilience against observed threats, not just hypothetical cases.

The discourse surrounding cybersecurity, particularly concerning critical infrastructure and Operational Technology (OT) systems, is often characterized by a stark dichotomy: either a state of pervasive, imminent catastrophe or an underestimation of genuine threats. Within this narrative, a crucial, yet frequently blurred, line exists between simulated exercises, vendor demonstrations of potential capabilities, and actual, successful breaches that have resulted in tangible disruption or damage. This distinction is not merely semantic; it is foundational to understanding the empirical reality of OT cybersecurity and informing effective policy and investment. The frequent conflation of "near-misses" and theoretical vulnerabilities with historical realities can lead to misallocation of resources, undue panic, and a skewed perception of risk. A rigorous examination is required to clarify the difference between a genuine failure and a demonstration of potential risk, underscoring the importance of empirical evidence over speculation.

Distinguishing Reality from Simulation

Cybersecurity vendor demonstrations and red team exercises, while valuable for illustrating potential threats, operate in controlled environments. A simulated attack against a test system, where defenders are aware of the exercise, bears little resemblance to a real-world breach. These exercises reveal potential vulnerabilities, not historical precedent; they show what *could* happen, not what *has* happened. The Aurora Generator Test is the cleanest illustration of why the line matters: a controlled laboratory experiment whose footage has been replayed for years as if it documented a real attack on the power grid.[24] The smoke was real; the attack was staged.

These demonstrations understate the complexity of a real-world attack, which involves overcoming unexpected network behaviors, diverse legacy hardware, and active security measures. Furthermore, real-world attackers are not necessarily motivated by the same objectives as a vendor showcasing a product. They may prioritize stealth for espionage, rapid financial gain through ransomware, or strategically timed disruption, and their methods will adapt to these goals, often involving prolonged reconnaissance and exploiting human or procedural weaknesses in addition to technical ones.

Similarly, penetration testing and "red teaming" exercises, when conducted properly, are invaluable tools for assessing an organization's defenses. These activities involve skilled professionals attempting to breach systems using methodologies that mimic real adversaries. However, the outcomes of these exercises are, by definition, controlled and contained. The participants are typically authorized, their objectives are defined, and the scope of their actions is delineated. A successful penetration test might demonstrate that a particular vulnerability exists and can be exploited, leading to unauthorized access or control. This is critically important for improving defenses. However, a red team's success should not be equated with a successful external attack causing widespread damage.

The very nature of a red team engagement means that the organization is aware of the exercise, has likely provided intelligence to the testers about the scope, and has established protocols for responding to a simulated breach. The absence of genuine catastrophe in such controlled scenarios, despite simulated success in gaining access, often points to the inherent resilience of OT systems and the effectiveness of layered security and operational safeguards. The fact that a red team can bypass certain defenses and gain access to a control network does not automatically imply that a nation-state actor or sophisticated criminal group could do the same in a live, unannounced attack, nor that they would achieve the same level of disruptive impact without detection. The critical difference lies in the element of surprise, the sustained operational tempo of a real adversary, and the potential for cascading failures in a complex, interconnected system that a red team might not be authorized or equipped to fully explore.

A related confusion sits one rung above the red team: the gap between an adversary's *access* and an adversary's *impact*. The Volt Typhoon warnings of 2023–24 are the live example. U.S. authorities documented a genuine, state-sponsored PRC campaign quietly pre-positioning on the IT networks of critical infrastructure, holding access for possible future use.[25] That is a real and serious finding. But access is not effect. As of this writing no outage, no physical damage, and no destructive action has been attributed to the campaign. To report pre-positioning as though the blackout had already occurred is to make the same category error as treating a red team's foothold as a catastrophe, with the added hazard that the actor here is real and the temptation to round "they are inside" up to "they have struck" is correspondingly stronger.

The distinction between demonstrating a capability and achieving a catastrophic outcome is also evident in discussions about attacks on water and wastewater systems. The Oldsmar, Florida, incident in 2021, where an individual was said to have remotely altered chemical levels at a water treatment facility, is often cited as an example of a successful cyberattack. But the narrative of "success" does not survive contact with the later record. Even on the original telling, vigilant oversight and engineered safeguards, a 24-to-36-hour lag before treated water would have reached customers and pH alarms positioned to fire long before harm, meant the public water supply was never in real jeopardy.[26] And the premise of an external attacker itself was never confirmed: the city manager later called the episode a "nonevent," and the FBI, per CyberScoop, could not confirm it was initiated by a targeted cyber intrusion at all.[27] Oldsmar served as a potent warning about exposed remote access and about the durability of an uncorroborated story, but it did not result in the feared catastrophic public-health crisis, and it may not have been an external attack at all.

Numerous other reports have emerged over the years about potential vulnerabilities in water systems, including access to SCADA networks and the potential to manipulate flow rates or chemical dosages. The Cyber Av3ngers activity of 2023–24 is the concrete recent instance, and it fits the pattern: internet-exposed Unitronics Vision-series PLCs reachable with the factory-default password "1111" on default port 20256 compromised and defaced with a political message, with no documented loss of water service.[28] While such reports point to the theoretical risk of contamination or operational failure, documented instances of actual, significant harm to public health or widespread service disruption stemming from cyberattacks remain exceedingly scarce. The sector's reliance on older, often air-gapped systems, coupled with robust physical safety mechanisms and operational staff trained to monitor for anomalies, has historically provided a degree of inherent resilience.

The manufacturing and heavy-industry sectors, particularly where processes involve hazardous materials, present the richest ground for separating near-miss from catastrophe, and the single most-cited "catastrophe" in this space is exactly where the evidence is thinnest. Hypothetical scenarios involving cyber-induced explosions or toxic releases are discussed constantly; the historical record offers, in the way of confirmed examples, principally the German steel mill incident disclosed in the BSI's 2014 annual report, whose entire public existence is one anonymous, undated paragraph in the BSI's annual report, naming no company, citing no malware, quantifying no damage, and unaccompanied by any public primary forensic report or malware sample in the decade-plus since.[29] That is not a basis for declaring the event imaginary; it is a demonstration of how slender the documentary foundation is beneath a very heavy claim. The complexity of industrial processes, the multiple layers of safety interlocks, and the fact that many manufacturing environments are not directly internet-connected contribute to a high degree of resilience. The incidents that do occur are dominated by ransomware that halts production lines, JBS, Toyota's supplier-driven shutdown, producing financial loss and downtime, or by data breaches that compromise intellectual property. Colonial Pipeline caused significant disruption, but it stemmed from ransomware on the IT and billing network, not a compromise of OT; the company took the line offline as a precaution and, in part, because it could no longer invoice.[30] The event underscored how IT compromises can *indirectly* halt OT operations, but it did not demonstrate direct cyber-physical manipulation of the pipeline's control systems for destructive purposes. The pipeline's OT systems were not themselves compromised by the ransomware; the shutdown was a precautionary measure taken by the company.

The distinction between a demonstrated vulnerability and a real-world catastrophic attack is crucial when evaluating the impact of state-sponsored cyber espionage on OT environments. Numerous reports highlight sophisticated actors, often attributed to nation-states, gaining access to sensitive industrial networks to steal intellectual property, gather intelligence on operational processes, or prepare for future disruptive actions. These activities are primarily focused on information acquisition rather than immediate physical destruction. Volt Typhoon belongs in this column too, insofar as the documented behavior is positioning and persistence rather than effect.[31] The long-term impact of such espionage and pre-positioning can be significant, affecting economic competitiveness and national security, but it does not equate to the immediate, widespread damage implied by the term 'catastrophe.' The absence of large-scale, physically destructive cyberattacks in the historical record, particularly within the United States, suggests that while the potential for such events is real, its actualization has been remarkably limited. This scarcity is likely attributable to a confluence of factors, including the inherent resilience of many OT systems, the significant technical hurdles and resources required for sophisticated cyber-physical attacks, the deterrent effect of attribution and potential retaliation, and the increasing effectiveness of defensive measures.

Therefore, when analyzing the historical record, it is imperative to maintain a clear distinction between different types of events:

- **Demonstrations and Tests:** These include vendor proofs-of-concept and authorized penetration tests. They illustrate potential vulnerabilities in controlled settings but are not actual historical breaches. a controlled demonstration at Idaho National Laboratory, is the archetype, and a reminder that the confusion is often manufactured, not accidental.

- **Near-Misses:** These are incidents where an attack is attempted or achieves initial access, but safeguards or human intervention prevent significant harm, *and* incidents whose alarming framing later proves unconfirmed. The Oldsmar water facility is the cautionary example on both counts: even the dramatic telling featured engineered delays and pH alarms that precluded harm, and the premise of an outside attacker was never confirmed, later being called a "nonevent" by the city's own manager.[32] TRITON also belongs here: a state-grade intrusion exposed precisely because the targeted safety system did its job and tripped the plant to safe.[33]

- **Successful but Non-Catastrophic Attacks:** These events cause significant operational disruption or financial loss without widespread physical destruction. The Colonial Pipeline ransomware attack, which forced a precautionary shutdown without ever touching the pipeline's OT, the ransomware-driven halt at JBS and the 2022 suspected-cyberattack shutdown at Toyota supplier Kojima Industries, the 2015–2016 Ukrainian grid outages, where manual operation kept customer outages to a matter of hours (though full SCADA restoration took months), and state-sponsored espionage and pre-positioning all fall into this category. They are the real, costly, instructive middle of the distribution, neither nothing nor apocalypse.

- **Catastrophic Cyber-Physical Attacks:** These attacks cause widespread, prolonged damage to critical infrastructure, resulting in mass casualties or a national security crisis. The public historical record contains no well-documented examples of an attack of this severity on U.S. critical infrastructure. Stuxnet, the one operation that achieved real physical destruction, did so abroad, against a single hardened nuclear-enrichment facility, at a cost few states could bear, and has not been replicated at scale in the public record.

This distinction is critical for an evidence-based approach to risk assessment. Focusing on the historical record allows for a pragmatic allocation of resources toward observed threats rather than hypothetical worst-case scenarios. The absence of past catastrophe does not guarantee future safety, but it provides vital context for protecting essential services. Nor does it license complacency: Volt Typhoon is a current, real, state-backed reminder that access is being sought and held, and the correct response to *access* is vigilance and hardening, not the rhetoric of a disaster that has not occurred.

The Role of Non-Adversarial Failures

Many OT disruptions originate not from malicious actors but from internal failures like human error, misconfiguration, or poor maintenance.

This perspective shifts focus from external adversaries to internal operational practices. Without strong protocols, human error remains a primary cause of disruption, where a single incorrect command or configuration error can cascade through automated systems. It is worth noting how directly this connects to the incident narratives above: the most credible alternative explanation for Oldsmar, offered eventually by the city's own manager, was not a foreign adversary but "the same employee... banging on his keyboard".[34] The line between a "cyberattack" and an internal mistake is, in practice, exactly as blurry as the attribution challenge that closes this chapter.

Numerous anecdotal reports and internal investigations within industrial sectors point to operational mistakes as common culprits. For instance, in the chemical processing industry, a miscalibrated sensor or an improperly set valve position, often a result of a maintenance error or an oversight during a procedural change, could lead to off-specification product, process instability, or even a safety shutdown. Similarly, in the power generation sector, an operator inadvertently opening a breaker at the wrong time, or misinterpreting a complex alarm sequence, can lead to localized outages or necessitate load shedding to maintain grid stability. These are not attacks; they are lapses in the human-machine interface, often exacerbated by fatigue, high-pressure environments, or a lack of comprehensive understanding of the system's intricate interdependencies. The learning curve for operating complex OT environments is steep, and the consequences of even minor deviations can be substantial. Without rigorous, ongoing training that includes realistic simulations of both normal and abnormal operations, and emphasizes the critical nature of precise input, the potential for accidental disruption remains high.

This is also where the strongest *steel-manning* of the alarmist case deserves a hearing. Engineers such as Joe Weiss have argued for years that the true count of OT incidents with physical consequences is higher than the public record reflects, because control-system environments frequently lack the forensic instrumentation that would let investigators tell a malicious manipulation from an equipment fault or a human slip.[35] The point is genuine and it cuts against complacency. The careful reader will notice, however, that it largely *reinforces* this section's theme rather than the doomsday one: the under-counted events Weiss describes are predominantly malfunctions, mis-operations, and insider errors, the non-adversarial failures catalogued here, not the remote-attacker-seizes-the-controls scenario that animates the fear narrative. The honest synthesis is that the murky middle is probably larger than we can see, and the spectacular top of the distribution remains as empty as the documented record suggests.

The challenges are compounded by the increasing complexity and interconnectedness of modern OT environments. Even in non-adversarial scenarios, interconnectedness amplifies the impact of errors. A mistake in a manufacturing plant's raw material handling system, for instance, might not only halt production but also trigger an automated shutdown in downstream processing units, leading to a plant-wide standstill. The rapid propagation of these effects, driven by automated interlocks and feedback loops, means that a localized human error can rapidly escalate into a significant operational incident. This is particularly true in environments that have undergone modernization, where previously isolated systems have been networked to improve data flow and enable remote monitoring and control. While these upgrades offer numerous benefits, they also inadvertently create pathways for errors to propagate more quickly and widely, the same IT-to-OT interconnection that turned Colonial's billing-system ransomware into a pipeline shutdown. Inadequate training and a lack of standardized operating procedures are recurring themes in the post-incident analyses of non-adversarial failures. In many OT environments, particularly in smaller organizations or those operating with legacy systems, there may be a reliance on a few highly experienced personnel who possess tacit knowledge of the system's quirks and operational nuances. While their expertise is invaluable, it also creates a single point of failure. If these individuals are unavailable, or if their knowledge is not effectively codified and disseminated through comprehensive training programs, the organization becomes vulnerable. New staff may be onboarded with insufficient practical guidance, leading to errors in execution. Furthermore, the pace of technological change in OT can outstrip the ability of training programs to keep up. As new functionalities are added or system architectures are updated, operators need to be retrained not only on the new features but also on how these changes interact with existing processes. A failure to provide this updated training can leave operators working with incomplete or outdated knowledge, increasing the likelihood of mistakes.

The phenomenon of "configuration drift" is another significant contributor to non-adversarial failures. Over time, systems are often modified to accommodate changes in production requirements, to implement minor upgrades, or to troubleshoot specific issues. If these changes are not meticulously documented and tested, the system's configuration can diverge from its intended state. This "drift" can introduce subtle instabilities or unintended dependencies that may not manifest until a particular operational condition is met or until a seemingly unrelated change is introduced. For example, an adjustment to a control loop's tuning parameters to improve efficiency under specific load conditions might, over time, compromise its stability under different, but common, operating loads. Without regular audits and validation of system configurations, these subtle changes can accumulate, creating a brittle system that is prone to failure under normal operational stress. This drift is often driven by the pressure to make quick, undocumented adjustments to maintain production.

System malfunctions are often a direct consequence of human choices in system design, resilience, or maintenance. Hardware failures, software bugs in supervisory or control applications, communication network issues, or even power fluctuations can lead to operational disruptions. While these are not directly attributable to human error in operation, they are often exacerbated by a lack of robust redundancy, inadequate diagnostic capabilities, or insufficient proactive maintenance. For instance, a cooling fan in a control cabinet failing might not immediately cause a system outage, but if it goes unnoticed, the subsequent overheating of critical components can lead to sporadic errors, performance degradation, and eventually, a complete system failure. This highlights the importance of comprehensive monitoring, alerting systems, and a proactive maintenance culture that addresses potential weaknesses before they cause tangible harm. The "if it ain't broke, don't fix it" mentality, while seemingly efficient in the short term, can lead to a gradual decay of system reliability, making it more susceptible to failure from minor internal or external stresses.

Incidents attributed to "cyberattacks" often have contributing factors related to human error or system malfunction. An attacker may exploit a vulnerability created by an internal lapse, such as an unpatched system or a default password left unchanged, the Unitronics PLCs that Cyber Av3ngers defaced were reachable precisely because the default "1111" had never been changed, and Colonial's intruders walked in through a single VPN account with no multi-factor authentication.[36] The boundary between "we were attacked" and "we left the door open" is, in such cases, almost entirely a matter of framing.

The complexity of OT security and incident response further complicates the attribution of causes. After a disruptive incident, organizations may blame external adversaries to avoid admitting internal failures, which can seem more damaging to their reputation. Oldsmar is the textbook case of how attractive the external-attacker story can be, and how long it can outlive the evidence for it.[37] This can lead to an incomplete understanding of the incident's root cause, hindering efforts to prevent recurrence. A thorough post-incident analysis must therefore be objective, looking beyond the immediate trigger to identify all contributing factors, including human, procedural, and technical elements.

The shift in focus from purely external threats to internal vulnerabilities also has significant implications for cybersecurity policy and investment. Addressing these internal risks requires investment in human capital, training, and operational protocols, including:

Comprehensive Training: Standardized, role-specific training for OT operators and engineers covering system operation, cybersecurity best practices, and incident response.

Standardized Operating Procedures (SOPs): Rigorously enforced, documented SOPs for all critical operations, with regular reviews and updates.

Human-Centered Design: Designing interfaces and workflows that minimize operator cognitive load and reduce the likelihood of errors.

Configuration Management: Robust processes to track all system changes, with regular audits to identify and correct configuration drift.

Proactive Maintenance and Monitoring: Shifting from reactive to proactive maintenance strategies, utilizing predictive maintenance techniques and continuous monitoring to identify and address potential hardware or software failures before they impact operations.

Culture of Reporting and Learning: Fostering an organizational culture where errors and near-misses can be reported without fear of reprisal, allowing for transparent analysis and learning. This "blameless postmortem" approach is crucial for continuous improvement, and it is the direct antidote to the reputational incentive that produces premature "we were hacked" narratives.

Layered Defense in Depth (Internal Focus): While external defenses are crucial, strengthening internal controls such as network segmentation, access controls, and anomaly detection within the OT network itself can mitigate the impact of human errors or system malfunctions. This includes ensuring that critical control functions are isolated and protected, even from internal network traffic that might be compromised or misconfigured.

By acknowledging the role of non-adversarial failures, organizations can build more resilient operational environments. True security requires mastering the internal complexities of human operation and system integrity, not just defending against external threats. Mitigating these internal risks is paramount to ensuring the reliable operation of critical infrastructure.

The Challenge of Attribution

Attackers can plant false evidence to mimic other adversaries, an ambiguity that allows states to shape narratives and complicates policy responses.

A primary technical hurdle is that attackers use sophisticated techniques to mask their origin. This includes the use of compromised intermediary systems (often referred to as "botnets" or "proxies"), anonymization services, and the exploitation of vulnerabilities that allow for the redirection of traffic or the manipulation of logs. Furthermore, advanced persistent threats (APTs) are designed for stealth and longevity, operating within networks for extended periods without detection. These actors may plant false flags to implicate other nations or groups, turning attribution into a high-stakes game where the evidence itself may be an illusion.

The incidents in this chapter show the full spectrum of how attribution actually behaves. At one end sits the disciplined, evidence-led attribution: the U.S. Treasury's 2020 sanctioning of the Russian state research institute TsNIIKhM over TRITON came with the weight of a formal government action and Mandiant's technical assessment behind it.[38] At the other end sits the *resistance* to overreach: in the Colonial Pipeline case, with a financially motivated criminal group as the obvious culprit, President Biden was careful to state in May 2021 that there was "so far no evidence" the Russian government was involved, declining the politically tempting leap from "Russian-based criminals" to "the Russian state".[39] And in the middle sits the cautionary Oldsmar example, where the most basic attribution question of all, *was there even an outside attacker?*, went from confident public assertion to an unconfirmed "nonevent" over the course of two years.[40] Together these show that attribution can be done responsibly, can be abused, and can simply be wrong, sometimes all within the same news cycle.

For instance, an attack might be crafted to resemble the "signature" of a known state-sponsored group, even if the true perpetrators are entirely different. This mimicry can serve multiple purposes: to deflect suspicion from themselves, to sow discord between the falsely implicated nation and its rivals, or to trigger a disproportionate response against an unintended target, thereby achieving a strategic objective without direct involvement.

Beyond technical challenges, the political dimension of attribution is equally significant in shaping public perception and policy. Without ironclad proof, attribution can become a political act driven by geopolitical rivalries, strategic interests, and domestic politics. Governments may attribute an incident to an adversary if it aligns with foreign policy objectives, even without conclusive evidence. This can lead to a situation where threat perceptions are amplified, and actions are taken based on speculation rather than certainty. The "politics of blame" can easily overshadow the pursuit of objective truth, creating an environment where accusations are readily made, but accountability is elusive.

Consider the historical landscape where attributions have been made with varying degrees of certainty, and in some cases, have been heavily contested. In instances where attribution has been swift and vocal, it often serves a political purpose. For example, following a significant cyber incident impacting a nation's infrastructure or economy, a government might quickly point fingers at a rival state. This immediate attribution can be used to galvanize public support, rally allies, and potentially justify retaliatory measures, whether diplomatic, economic, or even military. The supporting evidence may be circumstantial or rely on classified intelligence that is not publicly verifiable. The lack of transparency in intelligence gathering and analysis means that the public and even other governments are often left to accept attributions on faith, rather than through an open and rigorous evaluation of evidence.

Attribution ambiguity also allows actors to operate with impunity. If an attack can be sufficiently obfuscated, or if false flags are effectively deployed, the perpetrators may evade definitive identification. This lack of accountability can embolden them to continue their activities, knowing the risks of being identified are diminished. In the context of OT, where the potential for widespread disruption is immense, this evasion of responsibility is particularly concerning. An actor that successfully carries out a damaging attack on critical infrastructure and escapes attribution not only avoids immediate repercussions but also gains valuable experience and may even be perceived as a more formidable adversary, a perception that can itself be a strategic advantage.

Furthermore, the international legal framework for cyber warfare and attribution remains nascent and largely uncodified. While principles of international law, such as sovereignty and nonintervention, are often invoked, their application in the cyber domain is fraught with difficulty. The absence of universally agreed-upon definitions for what constitutes an "armed attack" in cyberspace, or clear protocols for evidence sharing and verification between nations, complicates efforts to establish collective attribution and response mechanisms. This legal and diplomatic vacuum allows for a greater reliance on unilateral attributions, which are often more susceptible to political influence.

The exploitation of attribution uncertainty is not limited to state actors. Sophisticated criminal organizations or even ideologically motivated groups can also benefit from the difficulties in pinpointing responsibility. By mimicking state-sponsored tactics or operating through proxies, they can achieve a similar level of deniability. This blurring of lines between state-sponsored attacks, criminal enterprises, and other malicious actors further complicates the attribution landscape. The Colonial Pipeline case is again instructive: a ransomware-as-a-service model deliberately decouples the malware's authors from its operators, so that even a confident technical link to "DarkSide" does not by itself establish who pulled the trigger or on whose behalf.[41] When an incident occurs, distinguishing between these different types of perpetrators, and indeed between a malicious attack and a non-adversarial failure, becomes a critical, yet often elusive, task.

The perception of threat is often driven by attribution, regardless of its factual basis. If a nation is consistently blamed for cyber incidents, even without definitive proof, its reputation can suffer, and it may face increased scrutiny, sanctions, or diplomatic isolation. This can have a chilling effect on international relations, fostering an atmosphere of distrust and suspicion. Conversely, states that are perceived as having robust cyber capabilities, and are also seen as responsible international actors, may find their actions subject to greater scrutiny when attributed with cyber incidents, necessitating a clear and evidence-based defense. The contested nature of attribution is evident in numerous public pronouncements and analyses. For example, following significant events, such as disruptions to power grids, financial systems, or critical infrastructure, government agencies may release reports detailing the incident and offering their assessment of responsibility. However, these assessments can sometimes be met with skepticism or outright denial from the accused party. In such cases, the public is presented with competing narratives, making it difficult to ascertain the objective truth. This information asymmetry, where one party may possess more detailed, albeit sometimes classified, information, can be strategically leveraged. The desire to attribute can also lead to a premature focus on "who" rather than "how" or "why." In the rush to assign blame, the opportunity to learn crucial lessons about vulnerabilities and methods may be lost. If the primary goal is political, the technical analysis needed to improve defenses is often overlooked. Moreover, the attribution process itself can be lengthy and resource-intensive, requiring specialized expertise in digital forensics, intelligence analysis, and international law. Not all nations possess these capabilities to the same degree, leading to an uneven playing field. This can result in situations where powerful states are more adept at making and enforcing attributions, while weaker states may find themselves as frequent targets of unsubstantiated accusations.

The impact of contested attribution extends to the realm of policy development. If there is a general lack of consensus on who is responsible for cyber incidents, it becomes challenging to establish effective international norms of behavior or to develop consistent and coordinated policy responses. This can lead to a fragmented approach, where individual nations act in their own perceived best interests, potentially exacerbating geopolitical tensions rather than resolving them.

In summary, the challenge of attribution combines technical complexity with political motivation. The difficulty in identifying perpetrators allows for deception, making it crucial to develop policies based on verifiable evidence rather than on the politics of blame. The incidents examined in this chapter, read once as they are told and once as the record supports, all point to the same modest, defensible discipline: describe what happened in the words the evidence will bear, distinguish access from impact and demonstration from disaster, and let the conspicuous scarcity of true catastrophe inform our spending and our fears in equal measure.

CHAPTER 3

THE MAZE OF U.S. GOVERNANCE: OVERLAPPING MANDATES AND PARALYSIS

U.S. cybersecurity governance for operational technology (OT) is a labyrinth, a complexity born not of design but of decades of crisis-driven evolution. As new threats emerged, new agencies and functions were created, distributing responsibility across an increasingly crowded landscape.

The result is a system of overlapping mandates that is often inefficient, confusing, and a central challenge to protecting the nation's critical infrastructure. No single statute laid out this structure; it accreted. Each agency in the roster that follows was added in response to a particular alarm, a particular incident, or a particular reorganization, and almost none of them were designed to fit together. The cumulative effect is the defining feature of American OT governance, and it is the feature this chapter sets out to document in full. The roster of agencies, the overlaps among their mandates, and the diffusion of accountability that results are the structural facts on which much of the rest of this book depends. Later chapters will return to this diagnosis rather than rebuild it; this is where it lives.

A Crowded Federal Landscape

At the federal level, responsibility is not singular but distributed across a staggering number of agencies.

The Department of Homeland Security (DHS) is a central figure, primarily through its Cybersecurity and Infrastructure Security Agency (CISA). CISA has a broad mandate to protect critical infrastructure, which includes disseminating best practices, providing technical assistance, and coordinating incident response.

However, CISA's authority is not absolute; its purview intersects with many other departments.

The Department of Commerce, through the National Institute of Standards and Technology (NIST), develops influential cybersecurity standards like its widely adopted Cybersecurity Framework. While NIST lacks enforcement authority, its frameworks are influential; other agencies often incorporate its recommendations into regulations, making them de facto requirements. NIST is, in this sense, the reference library of American cybersecurity, consulted by nearly everyone and binding on no one by its own hand. Its OT-specific guidance, NIST Special Publication 800-82 (Guide to Operational Technology (OT) Security), is the closest thing the country has to a common technical vocabulary for industrial control systems. [1] Yet whether any given utility must actually follow it depends entirely on which other agency, if any, has chosen to fold it into an enforceable rule.

The Department of Defense (DoD) and U.S. Cyber Command (USCYBERCOM) focus on defending military networks, but their mission extends to protecting critical infrastructure vital to national security. This blurs the lines between civilian and military domains, particularly against foreign adversaries. The required coordination between DHS and DoD highlights the complexity of agencies with different missions operating in the same threat space. A pipeline, a port, or a regional power authority can be at once a civilian commercial enterprise, a matter for a sector regulator, and an asset whose disruption a military planner regards as a national-security event. Each of those framings carries a different agency, a different legal authority, and a different reflex about what to do first.

Numerous other federal bodies have sector-specific responsibilities. The Federal Energy Regulatory Commission (FERC) regulates the electric grid, the Environmental Protection Agency (EPA) oversees water systems, the Federal Communications Commission (FCC) regulates communications, and the Securities and Exchange Commission (SEC) focuses on financial sector risk. The Department of Energy (DOE) serves as the lead sector-risk-management agency for the energy sector and houses its own specialized capability for the grid, while the Transportation Security Administration (TSA), an arm of DHS, carries authority over the security of pipelines and other transportation systems. [2] The result is that the energy sector alone answers, in different respects, to FERC, to DOE, to TSA for its pipelines, and to CISA as the national coordinator, before a single state regulator is even consulted.

Meanwhile, the Department of Justice (DOJ) and the FBI investigate and prosecute cybercrimes, a role that depends heavily on coordination with protective agencies like DHS. The FBI's instinct in a live intrusion is to treat the compromised system as a crime scene; the protective agencies' instinct is to treat it as a patient to be stabilized. Both instincts are legitimate. They are also, at the decisive early moment of an incident, frequently in tension, a tension this chapter will return to in concrete form.

This intricate web of federal agencies is further complicated by the roles of state and local governments.

The State and Local Dimension

A substantial share of critical infrastructure — commonly estimated at the majority of assets across most sectors — is privately held or managed at the state and local level. State governments often have their own emergency management agencies, public utility commissions, and cybersecurity task forces that work to protect infrastructure within their borders. They are crucial partners in implementing federal guidance, conducting local risk assessments, and responding to incidents that, while not national crises, can have devastating local impacts.

The National Governors Association, for example, has been a vocal advocate for stronger federal-state partnerships in cybersecurity. Many states have developed their own cybersecurity plans and initiatives, often mirroring federal efforts but adapted to their specific industrial bases and threat landscapes. The effectiveness of these state-level efforts can vary widely, depending on available resources, expertise, and political will. A state with a large investor-owned utility and a well-funded public utility commission may field a credible cyber program; a neighboring state whose water and power are run by dozens of small municipal authorities may have almost nothing. The map of American OT security is, in practice, a patchwork drawn not by threat but by tax base.

At the local level, municipal governments, county agencies, and special districts manage essential services like water, wastewater, transportation, and local power distribution. These entities are often the first responders to incidents and may have limited resources, making them particularly vulnerable. Their cybersecurity posture is frequently dependent on the guidance and support they receive from state and federal partners, but they also have direct operational responsibility. The water sector illustrates the scale of the problem: the country is served by a very large number of community water systems, the overwhelming majority of them small, locally governed, and operating with thin budgets and few if any dedicated security staff. [3] These are the entities at the bottom of the governance pyramid, and they are the ones holding direct operational responsibility for processes that, if disrupted, touch public health immediately.

The overlapping mandates are a consistent source of friction. CISA may work to protect the electric grid, while FERC sets mandatory standards, and the FBI investigates malicious activity. These complementary efforts require constant coordination to avoid duplicated work, conflicting guidance, or gaps in coverage. Each agency operates under its own legislative authority, budget, and reporting structure, creating a high potential for inefficiency. The point bears emphasizing because it is easy to mistake overlap for redundancy in the helpful sense, as if having many agencies meant having many backstops. In practice, overlapping authority without a single owner tends to produce the opposite of redundancy: it produces the assumption, held simultaneously by several parties, that someone else has the lead.

This jurisdictional complexity leads to several systemic issues. For critical infrastructure operators, it creates confusion over who to report to or which guidelines to follow, delaying incident response. The overlapping authorities can also fuel interagency rivalries, undermine collaboration, and prevent clear leadership during a crisis. Finally, with so many entities involved, responsibility is diffused. When an incident occurs, it is difficult to pinpoint failure, which slows learning and reform. Diffused accountability is not a side effect of this structure; it is the structure's most consequential output. A system in which everyone has a piece of the mandate is a system in which no one can be held cleanly responsible when the mandate is not met.

In essence, the U.S. cybersecurity oversight landscape is characterized by a pervasive lack of clear, singular authority and a high degree of interagency overlap. While efforts have been made to streamline and consolidate responsibilities, the sheer scope of critical infrastructure, the evolving nature of cyber threats, and the decentralized ownership of essential services mean that a complex, multi-stakeholder approach remains the reality.

This complexity, while potentially offering the advantage of diverse perspectives and specialized expertise, presents significant challenges in terms of coordination, efficiency, and accountability. For adversaries, it can represent an opportunity to exploit the seams and gaps inherent in such a distributed system. For those tasked with protecting the nation's critical infrastructure, it remains a perpetual challenge to navigate this maze effectively.

The Department of Homeland Security (DHS), established in the wake of the September 11, 2001 attacks, was envisioned as a unified entity to coordinate a wide array of federal agencies focused on preventing terrorism, securing borders, and responding to national emergencies. Within this sprawling department, the Cybersecurity and Infrastructure Security Agency (CISA) was formally established by the Cybersecurity and Infrastructure Security Agency Act of 2018, a pivotal moment intended to consolidate and elevate the nation's cybersecurity and infrastructure protection efforts under a single, empowered banner. CISA inherited a broad mandate: to lead the national effort to understand, manage, and reduce risks to critical infrastructure and cybersecurity. This includes sectors as diverse as energy, water and wastewater, communications, financial services, and transportation. The agency's responsibilities are multifaceted, encompassing the dissemination of threat intelligence, the provision of technical assistance and assessments, the development of cybersecurity best practices, and the coordination of incident response.

Despite the clarity of its mission statement, CISA's position within the broader DHS structure and its interactions with other government entities present a compelling case study in the challenges of consolidating authority within the U.S. governance maze. While CISA was designed to be the principal federal agency for cybersecurity, its operational effectiveness and ability to truly centralize oversight are continually tested by the inherent complexities of interagency dynamics and the departmental bureaucracy it operates within. The department itself is a vast ecosystem, housing numerous components with existing or overlapping responsibilities in areas that touch upon national security, emergency preparedness, and, by extension, cybersecurity. Agencies like the Transportation Security Administration (TSA), the Coast Guard, and even elements within Customs and Border Protection (CBP) have distinct but related missions that involve securing physical and cyber aspects of critical infrastructure. CISA must therefore navigate not only external interagency rivalries but also internal DHS dynamics to ensure a cohesive and efficient approach. It is worth pausing on the irony: CISA was created precisely to reduce fragmentation, and yet it sits inside a department that is itself a federation of components with their own histories, their own authorities, and their own claims on the cyber portfolio. Consolidation, in other words, was layered on top of fragmentation rather than substituted for it.

The effectiveness of CISA's mandate is also contingent on its ability to influence and coordinate with entities outside of DHS, many of which possess statutory authorities and regulatory powers that CISA itself lacks. For instance, when it comes to the energy sector, CISA can provide guidance and threat intelligence to utilities, but the Federal Energy Regulatory Commission (FERC) holds the direct regulatory authority to mandate cybersecurity standards for the bulk-power system. Similarly, in the telecommunications sector, the Federal Communications Commission (FCC) sets rules that impact network security. While CISA aims to be a central hub for information and best practices, its ability to enforce adoption of these practices often relies on the regulatory teeth of other agencies or the voluntary cooperation of private sector operators. This reliance on external authorities and voluntary compliance can dilute the impact of CISA's directives, making the consolidation of authority more aspirational than fully realized. The arrangement produces a peculiar division of labor in which the agency with the broadest mandate has some of the narrowest enforcement powers, while the agencies with real regulatory teeth each see only their own slice of the landscape.

Furthermore, covering all 16 critical infrastructure sectors designated under Presidential Policy Directive 21 (2013) and reaffirmed in NSM-22 (2023), presents a significant challenge. Each sector has unique operational characteristics, technological dependencies, and threat landscapes.

While CISA employs sector-specific liaisons and develops tailored programs, the depth of expertise and understanding required to effectively guide and support every sector can be immense. This can lead to situations where CISA's efforts are spread thinly, and its ability to provide deep, specialized support for Operational Technology (OT) security, which often involves legacy systems, unique protocols, and critical physical processes, becomes particularly challenging. OT cybersecurity demands a nuanced understanding that differs from traditional Information Technology (IT) security, requiring a specialized focus that can be difficult to maintain across such a broad portfolio. The engineer who understands the failure modes of a particular turbine governor or a particular chlorination controller is not interchangeable with a generalist who understands enterprise networks, and the pool of people fluent in both worlds is small. Spreading that scarce expertise across sixteen sectors, each with its own protocols and physics, is a structural constraint no amount of organizational will can fully overcome.

The creation of CISA was a direct response to a perceived need for a more unified and agile federal cybersecurity capability. Before its establishment, these functions resided primarily in the National Protection and Programs Directorate (NPPD), which the 2018 Act redesignated and restructured as CISA, alongside related functions distributed across other parts of DHS. The intention was to create a single point of contact for the nation's critical infrastructure operators and to streamline the federal government's response to cyber threats. However, the reality of its integration into the existing departmental and governmental framework means that CISA does not operate in a vacuum. Its success hinges on robust collaboration, information sharing, and a willingness of other agencies to cede or coordinate certain aspects of their cybersecurity roles. The persistent challenge lies in ensuring that CISA's lead role is recognized and respected across the entire government, and that its efforts are not fragmented by the independent operations of other agencies. Renaming and rechartering a directorate as an agency changed the org chart; it did not, by itself, change the incentives of every other department to guard its own turf.

Consider the process of developing and disseminating cybersecurity guidelines for OT. CISA, drawing on its threat intelligence and risk assessment capabilities, can identify vulnerabilities and recommend best practices. However, the National Institute of Standards and Technology (NIST), housed within the Department of Commerce, has a longstanding role in developing cybersecurity frameworks and standards that are widely adopted across industries. While CISA and NIST collaborate closely, the distinction between NIST's standard-setting function and CISA's operational and protective role can sometimes lead to overlapping guidance or confusion for stakeholders about which guidance takes precedence or is most applicable. CISA's role is often to encourage the adoption of NIST standards and other best practices, but the ultimate authority for mandating or enforcing these within specific sectors often rests with other agencies, such as FERC for the energy sector, or the EPA for water and wastewater systems. The challenge of consolidating authority is further illustrated by the persistent need for CISA to engage in extensive outreach and partnership-building. Its success is not measured solely by its internal capabilities but by its ability to foster trust and collaboration with hundreds of thousands of critical infrastructure asset owners and operators, as well as with state, local, tribal, and territorial governments. [4] This requires significant investment in relationship management and a constant effort to demonstrate value. When a significant cyber incident occurs, CISA is expected to lead the national response, but its ability to do so effectively depends on the seamless sharing of information and operational support from other federal agencies, including the FBI for investigative aspects, the intelligence community for threat attribution, and potentially the Department of Defense for specific national security implications. The internal structure of DHS, with its various operational components, can sometimes introduce its own set of coordination hurdles, requiring CISA to work through internal DHS channels as well as external ones.

Moreover, the funding and legislative mandates for CISA, while substantial, must also be considered in the context of the overall federal cybersecurity budget and the competing priorities of numerous agencies. The agency's ability to expand its workforce, invest in advanced tools, and conduct comprehensive research and development is subject to the annual appropriations process and the broader legislative agenda. This creates a dynamic worth naming plainly, because it sits at the heart of this book's larger argument. An agency whose budget is set annually, and justified annually, has a standing institutional interest in the salience of the threat it exists to counter. This is not a charge of bad faith; it is an observation about incentives. When the case for next year's appropriation is made in part by demonstrating the gravity and growth of the danger, the machinery of governance acquires a quiet bias toward emphasizing threat rather than reassurance, and toward expanding mandate rather than declaring a problem contained. The trajectory of CISA's own appropriations since its creation is the natural place to test this claim against the record. [5] The pattern of budget-driven mandate expansion is one to watch for across the entire roster of agencies described in this chapter, not CISA alone, and it recurs as a theme throughout the chapters that follow. Set against this, the same appropriations pressure can leave genuine needs underfunded: the critical need for specialized expertise in OT cybersecurity, including engineers and scientists who understand the intricacies of industrial control systems, may not always be adequately resourced if competing demands within DHS or other agencies are prioritized by lawmakers or department leadership. The budget can grow in aggregate while the specific, unglamorous capability the mission most requires goes wanting.

The ongoing evolution of the threat landscape also means that CISA's role and capabilities are in a perpetual state of adaptation. As new technologies emerge and adversary tactics become more sophisticated, the agency must continuously refine its strategies and resources. This dynamic environment makes the concept of permanently "consolidated authority" a moving target. What appears as a consolidated function at one point in time may need to be reevaluated and adjusted as the nature of cyber threats evolves, particularly those targeting the physical underpinnings of society through OT.

CISA's effectiveness, therefore, is a continuous negotiation, a constant effort to assert its leadership and coordinate a highly distributed and complex ecosystem of cybersecurity responsibilities across government and the private sector. The agency's existence represents a significant step towards streamlining, but the inherent fragmentation of U.S. governance means that achieving true, unimpeded consolidation of OT cybersecurity oversight remains an ongoing and formidable challenge.

Sector-Specific Silos

While CISA aims for centralization, much of the nation's cybersecurity oversight remains distributed among numerous sector-specific agencies. This fragmentation, intended to leverage domain expertise, often results in an inconsistent application of security standards and challenges in sharing threat intelligence.

This fragmented approach leads to inconsistent cybersecurity standards. Each sector-specific agency promotes its own guidelines, and while many draw from NIST frameworks, their application and enforcement vary widely. The energy sector has mandatory NERC reliability standards enforced by FERC. In contrast, the EPA's approach to the water sector relies more on voluntary guidance. This patchwork of oversight creates gaps, particularly where OT-specific threats are less understood or oversight is less stringent.

The water sector is the clearest case of a jurisdictional seam wide enough to fall through, and it is worth examining closely because it recurs in this book's later chapters. The contrast with the electric grid is instructive. The bulk-power system operates under the NERC Critical Infrastructure Protection standards, a body of mandatory, auditable, enforceable requirements with financial penalties behind them, administered through FERC. [6] A grid operator who ignores those requirements faces consequences. Water has nothing of the kind at comparable strength. The EPA is the sector-risk-management agency for water and wastewater, but its leverage over cybersecurity has rested largely on its existing public-health and drinking-water authorities rather than on any clean, purpose-built mandate to regulate the security of industrial control systems. When the EPA, in a March 2023 interpretive memorandum, attempted to read cybersecurity review into its oversight of public water systems through sanitary surveys, the effort drew legal challenge from state attorneys general and industry groups, was stayed by the Eighth Circuit, and was rescinded by EPA in October 2023, leaving the agency's cybersecurity reach over the sector unsettled. [7] CISA, for its part, can offer the water sector advice, assessments, and threat intelligence, but—pending finalization of CIRCIA reporting rules (proposed rule published April 2024; final rule pending as of 2025)—it cannot today compel a small utility to adopt any particular security control. The net result is a sector of public-health importance in which the federal government's protective agency has no teeth, the sector-risk agency's teeth are contested, and the FERC/NERC model of mandatory, enforceable standards that works for the grid has no analog in the water sector at all. The EPA-versus-CISA water gap is not a gap in attention. It is a gap in authority, and adversaries probe exactly such gaps.

This fragmentation also complicates cross-sector threat intelligence sharing. An adversary that targets the energy sector may use vulnerabilities present in manufacturing or transportation. The siloed nature of agency oversight, however, creates barriers to this intelligence flow. Agencies often hesitate to share sensitive information due to regulatory concerns or a lack of common reporting platforms, burdening CISA with bridging these informational gaps. A vulnerability discovered in a controller widely deployed across water, energy, and manufacturing is a single technical fact, but it can arrive at three different agencies through three different channels on three different timelines, if it arrives at all.

The Department of the Interior (DOI), for example, oversees sectors like water and power systems in certain regions, as well as natural resources critical to national infrastructure. Its cybersecurity responsibilities intersect with those of the EPA and the DOE, and its own internal structures for managing cybersecurity for its vast array of facilities and operational systems can be complex. This means that a threat identified within a DOI-managed hydropower facility might not be efficiently communicated or understood by the DOE's broader energy sector cybersecurity programs if interagency communication channels are not robust. The potential for a cascading failure, where a compromise in one sector enables a compromise in another, is a significant risk exacerbated by this distributed governance model.

Transportation presents its own version of the seam, and it raises a question of authority worth stating directly. Pipelines straddle the energy sector and the transportation sector, and security oversight of them falls not to FERC or DOE but to the Transportation Security Administration, an agency created and staffed primarily around the mission of securing aviation. For years TSA's pipeline-security work was conducted through voluntary guidelines rather than binding rules. When TSA issued a series of mandatory security directives to pipeline operators beginning in May 2021 (SD Pipeline-2021-01 and -02). [8] The directives were significant, but they also exposed a genuine question that the fragmented structure had left unanswered: does the agency built for airport checkpoints possess the institutional depth in industrial control systems to write prescriptive cybersecurity requirements for a continent-spanning pipeline network, and is a security directive issued under transportation authority the right instrument for the job? The early directives drew criticism from operators for being prescriptive in ways that did not always map onto operational reality, and were subsequently revised toward a more performance-based approach. [9] The TSA pipeline-directive episode is the fragmentation thesis in miniature: an authority assigned by the accident of organizational history rather than by technical fit, activated suddenly by an incident, and then forced to learn the domain in public.

The effectiveness of sector-specific agencies in promoting cybersecurity also depends on their relationship with the private sector entities that own and operate the vast majority of critical infrastructure. While these agencies are tasked with providing guidance and support, their ability to mandate compliance often varies. For instance, through security directives (such as the SD 1580/82-2022-01 series, first issued in October 2022 and subsequently revised), while the Department of Transportation's Federal Railroad Administration (FRA) issues safety regulations that can touch on cyber-physical concerns, but the extent to which these are proactively adopted and implemented by private rail companies can depend on a complex interaction of regulatory pressure, economic incentives, and the perceived threat landscape. CISA, in its role as a national coordinator, often relies on these sector-specific agencies to translate its broader guidance into actionable measures for industry stakeholders. When the sector-specific agency lacks sufficient authority, resources, or technical expertise in OT cybersecurity, this translation process can falter, leaving critical assets exposed.

Consider the example of cyber attacks targeting Programmable Logic Controllers (PLCs) or Supervisory Control and Data Acquisition (SCADA) systems. These are core components of OT infrastructure found across multiple sectors, including energy, manufacturing, and water. While CISA can identify threats to these systems generally, the specific impact and mitigation strategies might differ significantly. An attack on a PLC controlling a turbine in an electricity generation plant (DOE's purview) has different implications than an attack on a PLC managing a water treatment process (EPA's purview). The detailed technical knowledge and operational context needed to assess and respond to such attacks effectively are often held by the personnel within the sector-specific agencies. However, if the communication protocols for sharing incident details and lessons learned between the DOE, EPA, and CISA are inadequate, the insights gained from one incident cannot be effectively applied to prevent future occurrences in other, related sectors. This creates a reactive rather than proactive national defense. The same model of controller, exploited by the same technique, can be a documented lesson in one sector and an open door in another, simply because the two sectors answer to agencies that do not share a desk.

Legislative Paralysis: Why Governance Stays Broken

Beyond inefficient mandates, the political landscape itself creates policy paralysis, as partisan divides and powerful lobbying interests often derail or delay crucial cybersecurity initiatives.

Passing legislation for OT cybersecurity is notoriously difficult. Proposals to clarify agency roles, set standards, or provide resources often stall due to partisan disagreements over the scope of federal authority versus economic freedom.

Consider the debate over mandatory cybersecurity standards for private industry. Proponents argue voluntary measures are insufficient, while opponents cite economic burdens and government overreach. This ideological tug-of-war often results in legislative stalemates. A bill to empower CISA with regulatory authority will face opposition from industry groups fearing high costs, while a bill relying on voluntary frameworks will be criticized as too weak. The result is often weak legislation or a complete stalemate. The structural consequence is that the gaps documented earlier in this chapter, the water sector's contested authority, the pipeline regulator chosen by historical accident, the silence between the DOE and the EPA on a shared controller, are precisely the gaps that legislation could close and precisely the gaps the legislative process is least equipped to close, because closing them means assigning authority that some agency or some industry would rather not see assigned.

Lobbying efforts by various stakeholders further complicate this legislative landscape. Industry groups, representing sectors ranging from energy to finance to manufacturing, often invest significant resources in advocating for their members' interests. These efforts can be directed at shaping the language of proposed legislation, influencing committee deliberations, or even advocating for the outright defeat of certain bills. While lobbying is a legitimate part of the democratic process, it can lead to outcomes that prioritize short-term economic concerns over national cybersecurity. For example, a legislative push to standardize OT cybersecurity protocols across all critical infrastructure sectors might be met with resistance from companies that have invested heavily in proprietary systems or that fear the cost of upgrading to meet new, potentially burdensome, requirements. Lobbyists can effectively amplify these concerns, highlighting the potential economic disruption and influencing lawmakers to adopt a more cautious or less stringent approach, thereby contributing to policy inertia.

Bureaucratic resistance can also play a significant role in stifling legislative progress. Even if a bill makes it through committee and gains traction on the floor, entrenched interests within federal agencies themselves may subtly or overtly work against initiatives that could alter their established roles, responsibilities, or funding. Agencies that have long held sway over specific sectors might view attempts to centralize authority or introduce new oversight mechanisms as a threat to their autonomy and expertise. This can manifest in various ways, from providing cautious or even negative testimony during hearings to subtly influencing the drafting of legislation to preserve existing power structures. The complex interdependencies between agencies, as previously discussed, can be leveraged to create inertia; an agency may agree in principle to a proposed reform but then raise practical challenges or operational concerns that effectively bog down the implementation process, further delaying or diluting the legislative intent. The same incentive that biases an agency toward emphasizing threat at appropriations time, the instinct to protect and expand one's own mandate, operates in reverse when a reform threatens to redistribute that mandate to someone else.

The history of U.S. cybersecurity policy is replete with examples of promising legislative efforts that have ultimately faltered. For instance, attempts to establish a more unified federal approach to critical infrastructure cybersecurity, particularly in the aftermath of significant cyber incidents, have often faced an uphill battle. The debate over the creation of specific legal authorities for CISA to issue binding cybersecurity directives to critical infrastructure entities has been a recurring theme. While CISA has made strides in fostering collaboration and providing voluntary guidance, its ability to compel action, especially in sectors with strong existing regulatory frameworks or significant private sector opposition, has been limited. Legislative proposals designed to grant such authorities have frequently encountered the aforementioned roadblocks, partisan division over federal power, intense industry lobbying against perceived mandates, and interagency turf wars, resulting in stalled progress or the adoption of watered-down compromises. One of the few durable products of this period was a narrower achievement, a federal cyber-incident reporting mandate for critical infrastructure, whose long and contested path to enactment, and whose still-unfinished rulemaking, illustrate how even the most broadly supported reforms move at a pace measured in years. [10] This legislative gridlock directly impacts OT resilience. As policy lags behind technology, adversaries exploit vulnerabilities that government has failed to address. The lack of legally binding requirements creates a patchwork of security practices, where the weakest link can be exploited to the detriment of the entire system.

Furthermore, legislative inaction can create uncertainty for both industry and government. When the regulatory environment remains fluid or ill-defined, businesses may hesitate to make significant investments in cybersecurity, unsure of future requirements or the potential for retrospective mandates. This hesitancy can slow down the adoption of best practices and the implementation of advanced security technologies. Similarly, government agencies can find their efforts to coordinate and direct national cybersecurity strategies hampered by a lack of clear legislative authority. This ambiguity can lead to duplicated efforts, wasted resources, and a diminished capacity to respond effectively to large-scale cyber threats.

The issue is not merely one of partisan politics; it is also deeply intertwined with the structure of influence within the U.S. system. The intricate relationships between legislative committees, executive agencies, and private sector lobbyists create a complex ecosystem where policy can be shaped, delayed, or blocked by a relatively small number of key actors. Understanding these dynamics is crucial to comprehending why, despite the clear and present danger posed by cyber threats to OT, effective and comprehensive policy solutions often remain elusive. The legislative process, intended as a mechanism for deliberation and consensus-building, can, in practice, become a significant impediment to decisive action, particularly when dealing with the highly technical, rapidly evolving, and strategically critical domain of operational technology cybersecurity.

The resulting policy vacuum or inertia leaves the nation's critical infrastructure in a precarious position, vulnerable to adversaries who operate with far greater speed and agility than the legislative machinery of government can often muster. This highlights a fundamental challenge: how to reform or adapt a governance structure designed for a different era to effectively address the urgent, technology-driven security imperatives of the 21st century. The inertia born from legislative gridlock and political infighting is not a passive state; it is an active enabler of risk, allowing vulnerabilities to persist and deepen, with potentially catastrophic consequences for national security and economic stability.

The Tangible Costs of Confusion

This confusion has concrete consequences for the security of America's operational technology (OT). Fragmented governance causes delays in threat detection, response, and recovery, creating a dangerous gap between the speed of adversaries and the pace of government reaction.

Consider the process of incident detection. In a well-coordinated system, a potential intrusion into an industrial control system, perhaps a power grid substation or a water treatment plant, would trigger a cascade of notifications and analyses. Information would flow seamlessly between the affected entity, the sector-specific agency (SSA), the Cybersecurity and Infrastructure Security Agency (CISA), and potentially other intelligence and law enforcement bodies. In the current fragmented landscape, however, the initial alert might be mishandled because an agency with a partial mandate lacks the full visibility or technical expertise to interpret the threat. Critical early indicators could be missed, misclassified, or simply lost in the shuffle as different entities attempt to assert their jurisdiction or wait for definitive guidance from a higher authority. This delay in initial detection means that an adversary can gain a significant foothold, move laterally within the network, and achieve their objectives before the nation's defenders are even fully aware of an attack in progress. The window for containment shrinks, transforming a minor incident into a widespread disruption.

The problem is compounded during the response phase. Once a threat is identified, a coordinated and rapid response is paramount. This involves not only technical remediation but also clear communication, strategic decision-making, and potentially the mobilization of national resources. However, overlapping responsibilities and a lack of clear leadership mean that agencies often pursue parallel, uncoordinated response efforts. One agency might focus on attribution, another on containment, and yet another on public communication, all without a unified command structure. This leads to conflicting advice, redundant efforts, and a dangerous lack of clarity about who is in charge. For instance, during a critical infrastructure incident, it is essential to have a single point of contact that can authorize decisive action. In the absence of such a clear leader, decisions can be delayed as different agencies debate their roles, consult their legal counsel, or await consensus from a multitude of stakeholders. This bureaucratic quagmire allows attackers to prolong their presence within compromised systems, exfiltrate more data, or inflict greater damage.

The governance maze also creates significant obstacles during recovery, often the most prolonged phase of incident management. Restoring compromised OT systems to a secure and operational state requires deep technical knowledge of industrial processes, specialized equipment, and often, close collaboration between government and industry. When agencies are vying for control or when there is no clear lead for coordinating recovery efforts, the process can become excruciatingly slow. Industry partners may be unsure which government entity to work with for guidance on system restoration, cybersecurity best practices for reestablishment, or for any necessary federal assistance. This uncertainty can lead to a prolonged period of downtime for critical services, with cascading economic and societal impacts. A power outage that could have been limited to a few hours might stretch into days or weeks if the coordinated recovery effort is bogged down by interagency disputes or a lack of clear direction. The economic costs of such prolonged disruptions, including lost productivity and reputational damage, are immense.

These delays are not merely inconvenient; they create and exacerbate systemic vulnerabilities. The inefficiency inherent in the current governance structure means that critical infrastructure is not just vulnerable to sophisticated, nation-state-level attacks; it is also susceptible to less advanced, opportunistic intrusions, and even to accidental failures that could be averted with better coordination. The system's inability to respond swiftly and effectively to even moderately complex threats means that vulnerabilities, once identified, remain unaddressed for extended periods. This creates a persistent and pervasive risk environment where the nation's critical services operate under a constant, elevated threat posture.

This confusion is not theoretical. It manifests during real-world incidents. The scenario that follows is an illustrative composite, not a report of a single documented event: there is no water utility named Pineville, and the details are assembled from the kinds of jurisdictional collisions that the structure described in this chapter reliably produces. It is offered to make the mechanism legible, not as evidence in its own right. A book whose central argument is that the cybersecurity field too often substitutes vivid scenarios for documented fact owes the reader that distinction stated plainly, and the genuine, sourced incidents that anchor the argument appear elsewhere in these pages. With that caveat understood, the composite illustrates the governance failure with a clarity that the dry recitation of mandates cannot.

Vignette (a fictional composite): The Pineville Water Incident

When ransomware hit the Pineville Water District, a midsized utility, operators saw their control screens lock up one by one. The plant manager, following a state-level checklist, made three calls. The first was to the local FBI field office, which opened a criminal case and instructed him not to wipe any systems to preserve evidence. The second was to the CISA duty officer, who advised immediately disconnecting the affected network and restoring from clean backups to get the plant running. The third was to the regional EPA contact, whose primary concern was compliance with the Safe Drinking Water Act and who demanded an immediate plan for restoring service and a report on potential contamination, regardless of the cyber investigation.

Within two hours, the plant manager had three different federal agencies giving him three conflicting priorities: preserve evidence, rebuild the system, and restore service immediately. Paralyzed by the contradictory advice and fearing legal liability, he did nothing for six critical hours. The delay allowed the ransomware to spread further into the network, significantly complicating and prolonging the recovery.

Every element of that paralysis traces back to a real seam mapped earlier in this chapter. The FBI's evidence-preservation instinct, the protective agency's stabilize-the-patient instinct, and the EPA's public-health-compliance instinct are each, on its own terms, correct. What is missing is not competence at any one agency but an authority empowered to rank the three instincts and own the decision. The composite simply dramatizes, at the speed of a single shift, the diffusion of accountability that this chapter has argued is the defining output of the American governance structure. A real plant manager in a real small utility, facing real legal exposure, would have no better map than the fictional one did.

The lack of clear leadership and a singular, authoritative voice for OT cybersecurity further amplifies this systemic vulnerability. When there is no designated entity with the power to set clear, binding standards, or to enforce compliance, the adherence to cybersecurity best practices becomes inconsistent. Different organizations within the same critical infrastructure sector may adopt varying levels of security, based on their internal resources, risk appetite, or the limited guidance they receive. This creates a "weakest link" scenario where a single, less-secure entity can serve as an entry point for an attack that compromises an entire interconnected system. For example, a compromised network of a smaller, less-resourced transportation logistics provider could be used to disrupt the operations of larger, more secure entities within the same supply chain. The overarching governance structure fails to ensure a baseline level of security across the entire ecosystem, leaving it susceptible to the lowest common denominator.

Moreover, the lack of unified direction hinders the development and deployment of proactive defense strategies. Resources that could be used for threat intelligence sharing, vulnerability research, or the development of standardized security protocols are instead dissipated across multiple agencies, often with competing objectives. This fragmentation prevents the kind of holistic, national-level approach needed to anticipate and counter emerging threats. Instead of a cohesive strategy, the nation is left with a patchwork of initiatives, many of which may be redundant or even contradictory. This inefficiency is not just a theoretical problem; it translates directly into a reduced capacity to protect critical infrastructure from actors who are increasingly sophisticated and well-resourced. The absence of clear accountability further exacerbates these systemic vulnerabilities. When incidents occur, it can be difficult to pinpoint responsibility for failures in detection, response, or prevention. Was it a failure of a particular agency's intelligence gathering? Was it a lack of regulatory oversight? Or was it a consequence of insufficient legislative action? The diffusion of responsibility across numerous entities, coupled with the ambiguity of mandates, allows for blame to be deflected, making it challenging to learn from mistakes and implement necessary improvements. This lack of accountability creates a cycle of repeated failures, as the underlying governance issues that contributed to the incident remain unaddressed. The ultimate consequence is that critical infrastructure remains exposed to persistent vulnerabilities left unaddressed by bureaucratic inertia. The nation's OT systems are a fortress with many gates, each guarded by different patrols with unclear orders, a tangible result of a governance system that prioritizes process over protection.

This is the diagnosis on which later chapters build. When subsequent chapters reach for the phrase fragmented governance, or invoke the roster of overlapping agencies, or describe the seam between the EPA and CISA over water, or the question of TSA's pipeline authority, they are pointing back to the structural facts established here rather than deriving them anew. The maze described in this chapter is not a backdrop to the argument of this book. It is one of the conditions that allows a climate of inflated fear to flourish, because a system that cannot clearly assign responsibility for protection is also a system that struggles to speak with one calm, credible voice about what the real risks are and are not.

Chapter 4

THE VENDOR ECOSYSTEM: PROFITING FROM FEAR

The cybersecurity industry is a market driven by a fundamental imperative: revenue. But unlike markets for tangible goods, it thrives on the perception of invisible, complex, and potentially catastrophic threats.

This dynamic means the business of cybersecurity is not just about solving existing problems, but about cultivating a persistent sense of urgency regarding potential dangers. Here, a product's value is often tied less to its proven ability to prevent a breach and more to its capacity to address the *fear* of one.

This market, therefore, operates on a foundation of what can be termed "fear-as-a-service." Vendors and service providers are not simply selling tools; they are selling reassurance. They are selling the illusion of control in an inherently uncontrollable domain. This is achieved through marketing and sales strategies that emphasize the most alarming outcomes. The narrative is rarely about incremental improvements; instead, it focuses on existential threats to operations, intellectual property, and national security. The language used is often apocalyptic: "cyber warfare," "digital Armageddon," "unprecedented attacks." While sophisticated adversaries are real, the constant drumbeat of worst-case scenarios serves a commercial purpose: it keeps the perceived threat level high, justifying escalating investment in security products.

One does not have to take a critic's word for this. Some of the sharpest descriptions of the mechanism have come from inside the defensive establishment itself. In February 2017, Ian Levy, then the Technical Director of the United Kingdom's National Cyber Security Centre, the defensive arm of GCHQ, stood up and compared a large part of the threat-intelligence industry to "medieval witchcraft." Vendors, he argued, were dressing up ordinary criminal activity as the work of unstoppable supermen in order to sell protection against it. "If you call it an advanced persistent threat," Levy said, "you end up with a narrative that basically says 'you lot are too stupid to understand this and only I can possibly help you, buy my magic amulet and you'll be fine.' "[1] It is worth pausing on the source. This was not a disgruntled customer or an academic skeptic. It was a sitting government cyber-defense official, charged with protecting an entire nation's networks, telling the industry to its face that much of its marketing was a protection racket dressed in a lab coat. The "magic amulet" is the perfect emblem for fear-as-aservice: a talisman whose value lies not in any demonstrated power but in the dread it promises to ward off.

Consider the typical marketing collateral from a cybersecurity vendor. It will invariably feature stark statistics about the rising cost of data breaches, the increasing sophistication of malware, and the dire consequences of system downtime. Case studies, if presented, often focus on hypothetical scenarios or incidents that are sensationalized to highlight the vendor's purported solution. There is a deliberate emphasis on the "unknown unknowns", the threats that might emerge tomorrow, the zero-day exploits that could cripple an organization. This creates a powerful psychological effect. Organizations, particularly those responsible for critical infrastructure or sensitive data, are conditioned to believe that inaction or insufficient investment is a gamble they cannot afford to take. The potential cost of a single catastrophic breach, however unlikely, looms larger than the predictable cost of security products.

The security technologist Bruce Schneier gave this phenomenon its enduring name more than two decades ago. In his 2003 book *Beyond Fear*, he coined the term "security theater" to describe countermeasures that provide the *feeling* of security without meaningfully improving the reality of it.[2] Schneier's deeper observation is psychological, and it is the lever the entire fear market pulls on: human beings, he noted, systematically overestimate rare and spectacular risks while underestimating commonplace ones.[3] A spectacular, rare risk, a foreign state reaching through the internet to detonate a refinery, is precisely the kind of threat a frightened mind overweights, and precisely the kind a vendor is incentivized to dramatize. The commonplace risk, an unpatched server, a reused password, a misconfigured firewall, is the kind both the mind and the market discount, even though it is where the actual damage overwhelmingly originates. Markets do not reward selling the dull truth. They reward selling the feeling.

This dynamic is particularly potent within the operational technology (OT) sector. Unlike the more familiar realm of information technology (IT), OT systems are designed for longevity, reliability, and physical process control, not necessarily for rapid adaptation to evolving cyber threats. Legacy infrastructure, interconnected physical systems, and the devastating potential of cyber-physical attacks (e.g., power outages, water contamination, industrial accidents) make these organizations exceptionally vulnerable to fear-based marketing. A vendor can paint a compelling picture of a compromised power grid, a disrupted water supply, or a halted manufacturing line, and the inherent risks associated with these scenarios make the proposed solution seem indispensable. Abstract cyber threats are translated into concrete, terrifying possibilities, which in turn drives demand.

The business model is further reinforced by the inherent opacity of cyber-security efficacy. It is exceptionally difficult for a customer to independently verify the true effectiveness of a complex cybersecurity solution. When a breach *doesn't* happen, is it because the product worked, because no sophisticated attack occurred, or simply by chance? This ambiguity benefits the vendor. Success is rarely attributable, but the *potential* for failure is a potent motivator for purchase. The absence of an incident becomes tacit proof of the solution's value, a concept known as "negation by absence." Organizations pay for security that may not perform as advertised because the risk of *not* paying is framed as being far greater. Furthermore, the vendor ecosystem is adept at creating and disseminating narratives that solidify this market dynamic. Through industry conferences, white papers, analyst reports, and media partnerships, a consensus is often manufactured around the severity and imminence of particular threats. Cybersecurity firms frequently employ former government officials, intelligence operatives, and security researchers who lend an air of authority and insider knowledge to their pronouncements. These individuals can credibly articulate risks and the need for specific, often proprietary, solutions. Their pronouncements, amplified by the vendor's marketing, shape public perception and executive decision-making, creating a self-fulfilling prophecy of escalating threat and demand.

No career illustrates the journey from public warning to private profit more completely than that of General Keith Alexander. In July 2012, while serving as Director of the National Security Agency and head of U.S. Cyber Command, Alexander told an audience at the American Enterprise Institute that the ongoing theft of intellectual property through cyber means represented "the greatest transfer of wealth in history."[4] It was one of the most quoted lines in the history of the field, a phrase engineered to convert abstract risk into civilizational alarm. Two years later, in 2014, Alexander retired from government and founded a private cybersecurity company, IronNet Cybersecurity. He reportedly offered his consulting services to banks at rates of up to one million dollars per month.[5] The man who had told the country that adversaries were carrying out the greatest transfer of wealth in history was now, in effect, selling a private bulwark against that very transfer.

What happened next is the part of the story the fear narrative never tells, because it is the part where the fear and the profit detach from any underlying reality. In August 2021, IronNet went public through a merger with a special-purpose acquisition company at a valuation of roughly $1.2 billion.[6] The stock spiked to $31.26 a share.[7] Within months, the company slashed its annual recurring revenue projections by roughly sixty percent.[8] A subsequent shareholder lawsuit alleged that executives had misled the market about "transformational revenue" from deals, including a roughly $10 million arrangement with a Navy contractor and a roughly $22 million deal in Kuwait, that, according to the suit, "had no realistic chance of closing in FY22."[9] IronNet settled the litigation for $6.6 million.[10] The stock, which had touched $31.26, fell to twenty-six cents, was delisted from the NYSE in 2023, and the company filed for Chapter 11 bankruptcy in October 2023.[11] Bloomberg captured the arc in a single headline: "Former NSA Chief Keith Alexander Caught SPAC Fever, and Investors Got Burned."[12] The indictment writes itself. The general who warned the nation of "the greatest transfer of wealth in history" built a company on the strength of that warning, and the only large, documented transfer of wealth the company ever produced ran from retail investors into the pockets of insiders. The fear was real revenue. The defense was a balance sheet that collapsed to zero.

The concept of "threat intelligence" itself has become a significant profit center. While genuine, actionable intelligence is invaluable, the market is saturated with feeds and services that often provide generic, rehashed, or speculative information. The business imperative is to constantly provide *new* potential dangers, even if they are minor variations on old themes or threats unlikely to be faced by most organizations. The more threats are reported, the more solutions are perceived as necessary, creating a feedback loop that fuels demand for threat-detection products. This is the machinery Ian Levy was describing: an industry that profits in direct proportion to how frightened and how mystified its customers can be kept. The remedy he implicitly prescribed, demystification, is precisely what the business model resists, because a customer who understands the threat no longer needs to buy the amulet.

To counter this, consumers of threat intelligence should apply critical analysis. A simple checklist can help distinguish valuable intelligence from marketing material.

How to Read a Vendor Threat Report

1. **Check the Source:** Is the threat actor described with verifiable indicators (CVE numbers, hashes, IP addresses), or is it vague ("a sophisticated state-sponsored actor")?

2. **Assess Relevance:** Does the report clarify which industries or regions are at risk? A threat to Ukrainian energy companies may not be an immediate priority for a U.S. water utility.

3. **Distinguish Novelty from Repackaging:** Is this a new attack technique or a minor variation of a known threat presented as a novel crisis?

4. **Evaluate the Recommended Action:** Does the report suggest specific mitigations ("patch this CVE") or conclude with a generic recommendation to buy the vendor's product?

5. **Look for Independent Corroboration:** Have other reputable security firms or government agencies (like CISA) reported on the same threat? A lack of corroboration is a red flag.

The sales cycle in cybersecurity is also tailored to leverage this fear-driven market. Sales representatives are trained to probe for an organization's deepest security anxieties, to identify their most critical assets and their most plausible vulnerabilities, and then to present their company's offerings as the definitive bulwark against the imagined onslaught. The goal is not necessarily to educate the customer on nuanced risk, but to instill a sense of urgency and to present their solution as the only viable path to security. Discounts may be offered for immediate commitment, further incentivizing quick decisions driven by perceived immediate danger rather than thorough due diligence. The pressure to "buy now before it's too late" is a standard, if often unspoken, part of the sales pitch. It is Levy's "only I can possibly help you" rendered as a quarterly quota.

Moreover, the market is characterized by a proliferation of specialized solutions, each addressing a perceived niche threat or a specific layer of the IT/OT stack. This fragmentation can be strategically advantageous for vendors. It allows them to position themselves as experts in a particular domain, without necessarily having to prove the efficacy of their solution in a broader context. Organizations, faced with a bewildering array of options, may opt for multiple single-point solutions, believing that a layered defense is inherently superior, without a clear understanding of how these disparate tools integrate, or whether they are truly addressing their most significant risks. This leads to complex, expensive, and often inefficient security architectures, where the sum of the parts is less than the whole, but the revenue generated for the vendors is substantial.

The role of cybersecurity "experts" and consultants further solidifies this market. While many consultants provide invaluable, objective advice, the industry also includes those who are incentivized to identify problems that require their clients to purchase specific products or services. A consultant who recommends a comprehensive security overhaul, often involving significant investment in new technologies, is more likely to secure repeat business and generate higher fees than one who advises that an organization's current security posture is adequate, or that minor adjustments are sufficient. The discovery and articulation of potential vulnerabilities, whether real or exaggerated, become a valuable commodity.

The financial markets also play a role, and as the latter part of this chapter will argue, that role has grown from a supporting one into the engine of the entire system. Cybersecurity companies, particularly those that are publicly traded or seeking venture capital, are under immense pressure to demonstrate growth. This often translates into a strategic emphasis on expanding market share and increasing revenue, which can be more readily achieved by cultivating a sense of urgent need among potential customers. Investment analysts often track metrics such as "annual recurring revenue" and "customer acquisition cost," which incentivize vendors to focus on strategies that ensure a steady, predictable stream of income from a growing customer base. This economic reality encourages a focus on sustained demand, which is best generated by a continuous perception of threat.

The business of cybersecurity is a delicate balancing act. Vendors must convince customers of imminent danger without causing so much panic that they become paralyzed. The goal is a state of controlled anxiety that compels investment. This market is driven not by the absence of security, but by the *fear* of its absence. This perceived threat, cultivated and amplified, is the engine that powers the multi-billion-dollar cybersecurity industry, and the value of that engine, as we will see, is now counted not only in product sales but in the valuations placed on the fear itself.

The Theater of the Hack: Vendor Demonstrations

However, these demonstrations often disconnect from the everyday vulnerabilities of OT environments. While the notion of nation-state actors or highly sophisticated criminal enterprises targeting critical infrastructure is a legitimate concern, the daily reality for many OT operators involves threats that are far less glamorous but equally damaging. These can include simple phishing attacks that compromise an employee's credentials, misconfigurations in network security settings, unpatched legacy systems that have known exploits, or even insider threats stemming from disgruntled employees or accidental mistakes. The vendor demonstrations, in their pursuit of maximum impact, often focus on the "Hollywood" version of cyber-warfare, complex, multistage attacks that require significant resources and expertise to execute. While these extreme scenarios are possible, they can inadvertently overshadow the more mundane, yet pervasive, threats that are more likely to be exploited. This can lead to organizations investing heavily in solutions designed to defend against hypothetical, high-end adversaries, while neglecting basic security hygiene that would protect them from more common and accessible attack vectors. The focus shifts from practical risk mitigation to warding off an almost mythical, technologically advanced foe, the very "advanced persistent threat" Ian Levy warned was being used to make ordinary customers feel "too stupid to understand this."

Moreover, there is an inherent incentive for vendors to amplify the perceived severity of these simulated attacks. The business model necessitates the creation of a compelling narrative of urgency and existential threat. Consequently, demonstrations might be meticulously staged to ensure a successful outcome for the "attackers," highlighting the capabilities of the vendor's proposed defense. This can involve tailoring the simulation to bypass specific security measures that are known to be present but perhaps not optimally configured, or even creating scenarios that are technically feasible but highly improbable in a real-world context without a significant amount of prior reconnaissance and lateral movement within the target network. The line between a realistic simulation and an exaggerated performance can become blurred. The goal is not necessarily to provide an accurate representation of the most probable threats, but to present a scenario that is so alarming that the proposed solution appears indispensable. This practice, sometimes colloquially referred to as "fearware" marketing, leverages the psychological impact of perceived danger to drive sales. It is security theater in Schneier's precise sense, a performance engineered to produce a feeling. The technical prowess displayed in these simulations is undeniable, but its relevance to the everyday operational security posture of the target organization can be questionable.

The impact of these fearware-driven demonstrations on policy decisions and procurement processes is significant and often detrimental. When policy makers and procurement officers are primarily exposed to these high-fidelity, dramatic simulations, their understanding of OT cybersecurity risks can become skewed. Instead of focusing on a balanced risk assessment that considers the likelihood and impact of various threat types, decision-makers may become fixated on the most dramatic scenarios presented by vendors. This can lead to procurement decisions that are driven by the "wow" factor of a demonstration rather than by a thorough evaluation of the solution's efficacy against the organization's actual threat profile and existing vulnerabilities. Budgets may be allocated to advanced, expensive solutions designed to counter sophisticated APTs, while underfunding essential, yet less visually engaging, security measures like robust access controls, regular patching, security awareness training, and network segmentation. The pressure to be seen as proactive and technologically advanced can override the more pragmatic approach of addressing the most probable risks first.

This dynamic is further exacerbated by the inherent asymmetry of information in the vendor-customer relationship. Customers, particularly those within OT environments who may not have deep cybersecurity expertise, often rely heavily on the vendor's representation of threats and solutions. The technical complexity of both the attacks and the defensive mechanisms means that independent verification of a vendor's claims can be challenging and resource-intensive. When a vendor presents a compelling demonstration of their product preventing a simulated, devastating attack, it is difficult for the customer to objectively assess whether the threat was realistic, whether the vulnerability exploited was common, or whether the solution is truly as effective as portrayed. This reliance on vendor-provided narratives can lead to a situation where procurement is based on perception rather than objective performance. The vendor's demonstration becomes a powerful, albeit potentially misleading, piece of evidence in the decision-making process.

The procurement process itself can be influenced by these demonstrations. Requests for proposals (RFPs) may inadvertently include requirements that are directly derived from the capabilities showcased in vendor demonstrations, rather than from a genuine assessment of the organization's unique security needs. This can create a self-perpetuating cycle where vendors develop solutions to meet anticipated RFP requirements, which are in turn influenced by previous vendor demonstrations. The focus shifts from solving specific business problems to meeting generic, vendor-influenced technical specifications. This can lead to organizations acquiring expensive, highly specialized tools that are overkill for their needs, or that do not integrate well with their existing infrastructure, ultimately failing to provide the comprehensive security that was promised. The "fearware" demonstration, in this context, becomes not just a marketing tool, but a de facto standard that shapes procurement landscapes.

The increasing sophistication of OT environments, with their convergence of IT and OT, presents a new frontier for these vendor demonstrations. As OT systems become more interconnected and reliant on digital technologies, the potential attack surface expands, and the consequences of a breach become more severe. Vendors are quick to capitalize on this, demonstrating how cyber-attacks can traverse the IT-OT boundary, leading to physical disruptions. These demonstrations often highlight the specific vulnerabilities inherent in industrial control systems (ICS), such as legacy protocols, unencrypted communications, and limited patching capabilities. The visual impact of a simulated disruption to a chemical plant, a power grid, or a transportation network is exceptionally potent. It directly taps into the primal fear of real-world harm and societal disruption, making the proposed cybersecurity solutions appear not just as prudent investments, but as essential safeguards against societal collapse.

The role of third-party analysts and consulting firms, often engaged by organizations to guide their procurement decisions, can also be influenced by these demonstrations. If these analysts are themselves impressed by the technical prowess and dramatic impact of vendor-led simulations, their recommendations may lean towards solutions that excel in such demonstrations. This can perpetuate the cycle, as organizations trust their advisors to guide them towards the "best" solutions, which are often defined by their ability to impress in a simulated attack environment. Without independent, rigorous testing methodologies that focus on the actual effectiveness against common threats and the practical integration within specific operational environments, the influence of these fearware-driven demonstrations remains significant.

Ultimately, "fearware" demonstrations expose a fundamental tension in the market. Vendors have a legitimate interest in showcasing what their products can do; but the invisible nature of threats and the catastrophic consequences of failure create fertile ground for the exploitation of fear. When a dramatic simulation detaches from the reality of common vulnerabilities and is used to manufacture urgency, the result is misallocated resources and a market driven by anxiety rather than a rational assessment of risk. The true measure of a security solution is not its ability to impress in a staged attack, but its effectiveness against the actual threats an organization faces, day in and day out.

Of course, not all vendor activity is designed to exploit fear. The most effective vendors and consultants act as genuine partners, improving security outcomes through deep expertise and well-designed tools. These partners distinguish themselves by focusing on an organization's specific risk profile rather than generic doomsday scenarios. They provide transparent data on product efficacy, welcome independent testing, and build solutions based on open standards that avoid vendor lock-in. A trustworthy vendor helps an organization build sustainable, internal capabilities rather than fostering perpetual dependence on proprietary services. The key differentiator is a focus on measurable resilience over marketable fear.

The Black Box: A Lack of Independent Testing

The OT cybersecurity market lacks transparency regarding product efficacy. While vendors provide elaborate demonstrations, a critical vacuum exists for independent, rigorous testing and certification. This absence of standardized validation means the OT security landscape operates on perceived value rather than demonstrated resilience, leading organizations to invest based on marketing instead of proven effectiveness.

The challenge is multifaceted, stemming from the unique nature of OT environments and the inherent complexities of cybersecurity. Unlike consumer electronics or even many IT solutions, OT systems are designed for longevity, reliability, and continuous operation. This often means that security solutions must be integrated into existing infrastructure without disrupting critical processes, a task that is inherently more difficult and requires specialized knowledge. Furthermore, the potential consequences of a security failure in OT, ranging from plant shutdowns and environmental damage to loss of life, are far more severe than in traditional IT settings. This heightened risk profile necessitates a higher degree of assurance in the security products deployed. However, the mechanisms for providing such assurance are often underdeveloped or entirely absent.

This opacity stems from the scarcity of independent testing frameworks for OT. General IT security certifications fail to address the distinct protocols and architectures of industrial control systems (ICS). A solution that works in an IT network might fail in a real-world substation due to proprietary protocols, legacy systems, and physical process dependencies.

Vendors are thus left in a position where they can, with relative ease, make ambitious claims about their product's ability to detect, prevent, or mitigate specific OT threats. These claims might be based on internal testing, theoretical models, or demonstrations that, as previously discussed, can be meticulously curated to showcase the vendor's strengths. However, without an independent body rigorously subjecting these products to a standardized battery of tests that mimic realistic OT attack scenarios, including edge cases and common miscon-figurations, potential customers are left to take these claims at face value. This creates an uneven playing field where innovative yet potentially less market-savvy vendors offering truly robust solutions may struggle to compete against those with more sophisticated marketing departments and demonstration capabilities, even if their products are less effective in practice.

The economic implications of this lack of transparency are significant. Organizations, particularly small and medium-sized enterprises (SMEs) or those within critical infrastructure sectors with limited cybersecurity expertise, often find themselves vulnerable to what can be termed "security theater." They invest heavily in solutions that appear impressive in demonstrations or are heavily marketed as the "best" by vendors, only to discover later that these solutions do not provide the actual protection they were led to believe. This wastes financial resources and, more critically, creates a false sense of security that makes them more susceptible to breaches. The budget for foundational practices like network segmentation, patching, and access control is instead spent on a solution that may offer little benefit against the most likely threats.

Consider the scenario of an industrial facility seeking to secure its operational network. A vendor might present a sophisticated intrusion detection system (IDS) that claims to detect zero-day exploits targeting specific industrial protocols. The demonstration might be compelling, showing the system flagging a simulated advanced persistent threat (APT) with impressive speed and accuracy. However, what the demonstration likely omits is the system's performance in identifying more common threats, such as an employee accidentally downloading malware via a phishing email that then attempts to spread laterally within the OT network, or a misconfigured firewall rule allowing unauthorized remote access. Furthermore, the process of tuning such an IDS in a live OT environment can be exceptionally complex and time-consuming, often requiring specialized expertise that the customer may not possess. If the vendor's support package does not adequately address this, the sophisticated IDS may end up generating a high volume of false positives or, worse, failing to detect the most probable threats.

The reliance on vendor-driven testing also creates a perverse incentive. Vendors are naturally inclined to test their products in environments and against scenarios that highlight their strengths and minimize their weaknesses. This can lead to the creation of "golden test environments" that do not accurately reflect the messy, heterogeneous, and often legacy-laden reality of real-world OT deployments. Moreover, the specific threats chosen for testing may be those that are easiest to detect with their particular technology, while overlooking more insidious or subtle attack vectors. This self-testing paradigm is like a student grading their own exam; the results offer little objective measure of true capability. The absence of readily available, comprehensive efficacy data also hinders the development of best practices and informed policy. When government agencies or industry standards bodies attempt to formulate guidelines or mandates for OT cybersecurity, they often lack the independent data needed to make evidence-based recommendations. They are, in turn, susceptible to the narratives and claims put forth by the vendor community. This can lead to policies that are either too vague to be effective or that inadvertently favor certain types of solutions championed by dominant vendors, rather than those that are demonstrably most effective across the board. The procurement process becomes less about security and more about compliance with potentially flawed or vendor-influenced standards.

The challenge is compounded by the high cost of robust, independent testing, which is itself a barrier to entry for smaller vendors or those with niche solutions. Only larger companies with substantial resources can afford to submit their products to rigorous, multifaceted testing and certification processes. This entrenches the dominance of a few key players who may have prioritized marketing over deep, verifiable efficacy, and risks a market where the loudest voices and the most visually impressive demonstrations win, rather than where the most effective and resilient solutions prevail. Compounding this, the lack of standardized metrics makes it difficult for organizations to compare solutions at all: vendors report detection rates, response times, and mitigation capabilities using proprietary measures that are not directly comparable to those of their competitors, forcing customers to rely on testimonials and demonstrations and reducing procurement to educated guesswork rather than data-driven analysis.

The problem is not necessarily that vendors are intentionally deceiving their customers. Many vendors genuinely believe in the efficacy of their products and are proud of their capabilities. However, the market dynamics, coupled with the inherent complexities of OT cybersecurity, create an environment where unsubstantiated claims can thrive, and where the absence of rigorous independent validation becomes a significant vulnerability in itself. This vulnerability affects not only individual organizations but the resilience of critical infrastructure and national security as a whole. As OT systems become increasingly interconnected and digitized, ensuring the verifiable effectiveness of the cybersecurity solutions deployed is no longer a matter of good practice, but an imperative for safety, security, and societal stability. Addressing this opacity requires a concerted effort from all stakeholders to establish clear, verifiable performance standards and foster a market where proven efficacy, not compelling marketing, is the ultimate currency.

The Gilded Cage: Vendor Lock-In as a Strategic Risk

The proprietary nature of many OT security solutions fosters vendor lock-in, a significant strategic vulnerability. When an organization relies on a single vendor for its OT cybersecurity, the initial investment in hardware, software, and training creates powerful inertia. The perceived cost and complexity of migrating away from this established ecosystem, often involving significant re-engineering of networks, retraining of personnel, and the procurement of entirely new suites of products, can seem insurmountable. This entrenchment, while often driven by a desire for seamless integration and vendor support, can paradoxically undermine the very resilience it aims to achieve.

Consider a large-scale energy utility that has implemented a comprehensive security monitoring system from a particular vendor across its grid management facilities and remote substations. This system likely involves proprietary sensors, specialized software agents installed on industrial control systems (ICS), and a centralized analysis platform. The initial deployment would have necessitated extensive consultation, custom configuration, and potentially bespoke integration with existing SCADA protocols. Years down the line, if this vendor's technology begins to lag behind emerging threats, or if the vendor itself faces financial instability or is acquired by a less scrupulous entity, the utility finds itself in a precarious position. The cost of replacing the entire system, including operational shutdowns, retraining, and ensuring interoperability with legacy components, could run into the tens of millions. This financial disincentive forces the utility to rely on a potentially suboptimal solution.

The "acquired by a less scrupulous entity" scenario is not hypothetical, and to see why, one only has to follow the ownership of the OT-security industry itself. A striking and under-examined feature of the operational-technology security market is that the companies selling protection against industrial cyber threats are increasingly owned by, or funded by, the very firms that build the industrial equipment in the first place. The threat narrative and the equipment vendor have become, in a growing number of cases, the same balance sheet. Schneider Electric and Honeywell, two of the largest makers of industrial control systems on earth, have both invested in OT-security startups — Schneider Electric in both Claroty and Nozomi Networks, Honeywell in Nozomi.[13] In 2025, Mitsubishi Electric, another major industrial-equipment manufacturer, agreed in 2025 to acquire Nozomi Networks outright, completing the deal in January 2026, at a reported value of roughly one billion dollars.[14] In 2023, Rockwell Automation, one of the dominant suppliers of programmable logic controllers and factory-automation gear, acquired the OT-security firm Verve Industrial.[15] That same year, Honeywell acquired SCADAfence, folding another OT-security vendor into an industrial conglomerate.[16] And the OT-security firm Dragos has counted among its backers the venture arms of Koch Industries and Saudi Aramco.[17]

Sit with the structure of that arrangement. The same corporate families that sell you a control system with a decades-long lifecycle, built in an era with little regard for security, now also sell you, or own the firm that sells you, the monitoring platform that watches that control system for attacks. The equipment maker has a commercial interest in your continued anxiety about the equipment it sold you, because that anxiety now feeds a second revenue line on the same income statement. This is the gilded cage with the door welded shut: the incentive to characterize the installed base as dangerous and the incentive to sell the danger-monitoring sit inside the same enterprise. Independent assessment of whether the threat is real, or whether the original equipment should simply have been built more securely, is the last thing such an arrangement is structured to produce.

This dependency extends beyond mere financial considerations; it impacts the agility and adaptability of critical infrastructure. In the rapidly evolving landscape of cyber threats, the ability to quickly adopt new security paradigms and patch emergent vulnerabilities is paramount. Vendor lock-in can cripple this adaptability. If a new, more effective intrusion detection technique or a novel encryption standard emerges, but it is not compatible with the incumbent vendor's ecosystem, adoption becomes a Herculean task. This can leave critical systems exposed to known, preventable attacks simply because the chosen vendor's roadmap does not align with industry advancements or the specific security needs of the operator. The commercial relationship, intended to enhance security, transforms into a straitjacket, preventing the adoption of more robust and forward-looking solutions.

The implications for national security are particularly stark. Critical infrastructure, power grids, water treatment plants, transportation networks, and petrochemical facilities, forms the backbone of a nation's economy and its ability to function. If these vital systems are reliant on proprietary technologies from a limited number of vendors, a nation becomes vulnerable to disruptions orchestrated by adversaries targeting those specific vendors. If a foreign state actor compromises a key supplier of ICS software, the difficulty of transitioning away from that vendor's products creates a persistent, systemic risk. The adversary gains leverage, knowing a future attack could cripple essential services because the affected entities cannot rapidly divest from the compromised technology. This creates a scenario where economic pressure, espionage, or even direct cyberattacks on the vendor can have cascading effects on national security. The reliance on a single vendor for security also concentrates risk. If that vendor's internal security practices are lax, or if their product contains a deeply embedded, unknown vulnerability, the entire ecosystem built upon it is compromised. Unlike a diverse, multi-vendor environment where a breach might be contained, a compromised central vendor platform can provide a direct ingress point into numerous critical facilities. A vendor's backdoors for support or diagnostics, if exploited, can become catastrophic points of failure. The convenience of an integrated vendor solution can become its Achilles' heel.

Moreover, the long lifecycles of OT systems exacerbate the vendor lock-in problem. Industrial equipment is often designed to operate for decades, far longer than typical IT hardware. This means that a decision made today regarding a cybersecurity vendor could lock an organization into a specific technology path for twenty or thirty years. During this extended period, the geopolitical landscape can shift dramatically, new technological paradigms can emerge, and the vendor's own business strategy or ownership may change, potentially introducing new risks or misalignments with national security interests. A vendor that is a trusted partner today might, through acquisition or changes in national allegiance, become a source of concern tomorrow. The inability to easily pivot away from such a vendor due to the entrenched ecosystem can leave critical infrastructure exposed to evolving geopolitical threats. And as the ownership map above makes plain, the cybersecurity industry itself is in a state of perpetual acquisition, so the question of who will own any given OT-security platform a decade from now is genuinely open.

The commercial pressures that drive vendor lock-in are often disguised as a commitment to support and specialization. Vendors frequently argue that their deep understanding of a specific product line, coupled with their specialized training programs, offers superior security assurance. They may offer lucrative long-term support contracts, promising round-the-clock assistance and guaranteed updates. While this can be appealing, especially for organizations with limited in-house cybersecurity expertise, it reinforces the dependency. The cost of these support contracts can be substantial, and the terms often include clauses that penalize early termination, further increasing the financial burden of disentanglement. This creates a perpetual revenue stream for the vendor, funded by the very dependency they have cultivated, and it is precisely that perpetual, predictable revenue stream that makes these companies so attractive to the financial buyers we will turn to shortly.

Interoperability, a cornerstone of modern IT and increasingly a necessity in OT, is frequently sacrificed at the altar of vendor-specific solutions. Proprietary protocols and closed architectures, common features of locked-in ecosystems, hinder the ability of different security tools to communicate and collaborate effectively. This fragmentation can lead to blind spots in an organization's overall security posture. For instance, a specialized OT intrusion detection platform might not be able to share threat intelligence with a broader network security monitoring system from a different vendor, or with an endpoint detection and response solution from a third. This inability to share data means a sophisticated, multilayer attack might go undetected. The vendor's promise of an all-in-one solution, while attractive, can lead to a less effective security architecture. The irony is sharp: the consolidation of the industry into ever-larger ownership groups is justified, in the investor decks, as a path toward integrated "platforms," even as the underlying products remain proprietary islands that do not readily talk to a competitor's tools.

The narrative around vendor lock-in also often overlooks the potential for obsolescence. Technology moves at a relentless pace. A vendor's product that is cutting-edge today may be legacy technology in five to ten years. If an organization is locked into a vendor that is slow to innovate or whose product roadmap no longer aligns with industry trends, it risks deploying cybersecurity measures that are increasingly ineffective against emerging threats. This is particularly dangerous in OT, where the long operational life of assets means that a security solution might need to remain functional and effective for decades. Without the ability to refresh or replace components of the security architecture as technology advances, critical infrastructure can become vulnerable to attacks that exploit outdated technologies, a scenario that is often simpler and more predictable for adversaries to execute.

Addressing the vendor lock-in effect requires a multi-pronged approach. Governments and regulatory bodies play a crucial role in promoting open standards and mandating interoperability requirements for security solutions deployed in critical infrastructure. Procurement policies should actively favor solutions that adhere to open protocols and offer modular designs, allowing for easier component replacement and integration with third-party tools. Incentivizing the development of robust, independent testing and certification programs, as discussed in previous sections, can also help mitigate the reliance on vendor-provided assurances. By providing objective data on product efficacy and compatibility, these programs empower asset owners to make more informed decisions, reducing the power of proprietary ecosystems.

Furthermore, educating OT asset owners about the risks of vendor lock-in and the benefits of adopting a multi-vendor, standards-based approach is essential. This includes promoting awareness of the total cost of ownership, which extends beyond initial purchase price to include long-term support, integration challenges, and the potential costs of future migration. Encouraging the development of in-house expertise, or fostering partnerships with independent system integrators who can design and manage heterogeneous security environments, can reduce the perceived need for a single, all-encompassing vendor solution. Ultimately, a resilient critical infrastructure cybersecurity posture is one that is adaptable, interoperable, and not beholden to the strategic interests or technological limitations of a single commercial entity. The transition away from deeply entrenched vendor ecosystems is challenging, but it is a necessary step to ensure the long-term security and stability of the nation's vital assets.

Shaping the Rules: Vendor Influence on Policy and Standards

The cybersecurity vendor ecosystem exerts a complex and often opaque influence over the policies and standards designed to govern it. This influence is wielded not through coercion, but through lobbying, engagement in policy forums, and the strategic deployment of industry expertise. The result can be a regulatory and standards landscape that, intentionally or unintentionally, aligns closely with the commercial interests of the vendors themselves, potentially creating an environment where proprietary solutions are favored, and broader systemic governance improvements are sidelined.

A significant avenue for vendor influence lies in direct lobbying. Large cybersecurity companies engage policymakers to shape the narrative around threats, steering the conversation toward areas where their products offer a solution. For instance, a vendor specializing in industrial control system (ICS) anomaly detection might lobby for policies that mandate its specific monitoring technologies. This lobbying is often framed as providing essential expertise, with vendors presenting themselves as indispensable partners in national security.

Crucially, this influence does not require a smoke-filled room. The most powerful lever vendors possess is the regulatory ratchet itself. When government mandates require organizations to take specific security actions, every one of those mandates converts directly into demand for products. Consider two recent and concrete examples. In 2023, the U.S. Securities and Exchange Commission adopted rules requiring public companies to disclose material cybersecurity incidents within four business days of determining that they are material.[18] In the European Union, the NIS2 Directive dramatically expanded the universe of "essential" and "important" entities subject to mandatory cybersecurity risk-management and incident-reporting obligations, sweeping in tens of thousands of additional organizations across critical sectors.[19] Each rule is, on its face, a reasonable response to genuine risk. But each rule also instantly manufactures a market. A four-day disclosure clock requires detection, forensics, and legal-readiness tooling that a company must buy. An expanded roster of regulated entities under NIS2 requires every newly covered organization to procure risk-management and reporting capabilities it did not previously need. The fear produces the narrative; the narrative produces the regulation; the regulation produces the mandated spend. And the firms best positioned to satisfy each new mandate are, very often, the same firms whose lobbying and expert testimony helped shape its contours.

Beyond direct lobbying, vendors actively participate in the formation of cybersecurity standards. Numerous government agencies and industry consortia establish committees and working groups dedicated to developing best practices, technical standards, and frameworks for cybersecurity, particularly within critical infrastructure. Representatives from leading cybersecurity firms are frequently appointed to these bodies, bringing with them deep technical knowledge and a keen understanding of their own product capabilities. While their contributions are often invaluable in providing practical, ground-level insights, their inherent commercial interests can subtly shape the outcomes. Standards that require specific types of data collection, reporting formats, or even particular technical architectures can inadvertently create a market advantage for vendors whose products are designed to meet those precise specifications. This can manifest as a preference for prescriptive standards, detailing *how* a control must be implemented rather than *what* outcome is required.

Effective security demands a shift from prescriptive standards, which dictate *how* a control is implemented, to performance-based standards that define *what* outcome is required. This shift fosters more diverse and adaptable solutions.

During the development of frameworks like the NIST Cybersecurity Framework, vendors engage in public comment periods to advocate for specific language or risk mitigation strategies that align with their products. For example, a vendor offering endpoint detection and response (EDR) solutions might push for standards that emphasize granular endpoint visibility, a requirement that naturally benefits its products. While not inherently malicious, this influence can predispose frameworks to favor certain vendor solutions, hindering the adoption of innovative or cost-effective approaches from smaller competitors.

The 'revolving door' phenomenon, the movement of individuals between government and private sector cybersecurity roles, also shapes policy. Professionals who have served in government roles often possess valuable insights into policy processes, regulatory requirements, and government procurement strategies. Conversely, former industry executives and technical experts bring their deep product knowledge and market understanding to government roles. Powerful networks of influence are forged. A former government official now working for a cybersecurity vendor may leverage their established relationships and understanding of policy to advocate for their new employer's interests. The arc of Keith Alexander, from the directorship of the NSA to the founding of the cybersecurity firm IronNet, which sold protection against the very threats he had spent his government career describing, is the revolving door in its purest form, and it is far from unique. Similarly, former industry insiders within government agencies may possess an inherent bias towards solutions and approaches with which they are familiar from their private sector experience. This dynamic leads to procurement decisions and policy directives that are heavily shaped by the perspectives and commercial objectives of the vendors.

Moreover, the sheer complexity and rapid evolution of cybersecurity threats often create a knowledge asymmetry between vendors and government bodies. Vendors, by virtue of their continuous engagement with emerging threats and their focus on developing defensive technologies, often possess a more current and granular understanding of the threat landscape and the efficacy of various mitigation techniques. This technical expertise makes them sought-after advisors. When this advisory role becomes dominant, policy can become reactive to the solutions vendors offer, rather than proactively shaping a more resilient security posture. The urgency of cybersecurity, coupled with this expertise gap, often leads policymakers to rely heavily on vendor recommendations, sometimes without sufficient independent verification.

The commercial incentive for vendors to shape policy is substantial. Government procurement, especially for critical infrastructure and national security, represents a massive market. By influencing standards and policies, vendors can effectively "pre-qualify" their products and services, creating a more predictable and lucrative sales pipeline. They can lobby for mandates that require specific functionalities, thereby creating demand for their solutions. This can also extend to setting requirements for certifications and accreditations, where vendor-developed testing methodologies or proprietary frameworks might be subtly integrated into government processes, giving an edge to companies that have already invested in these areas.

Furthermore, the focus on specific product categories by vendors can narrow the scope of policy discussions. If the dominant vendors in a particular sector are focused on areas like network intrusion detection, endpoint security, or data encryption, policy discussions may gravitate towards these technologies. This can lead to underinvestment or neglect in other critical areas of cybersecurity, such as secure software development practices, supply chain security, or foundational principles of operational resilience that are not tied to a specific vendor product. The emphasis shifts from holistic risk management to the deployment of discrete, often vendor-provided, security tools.

The narrative presented by vendors often emphasizes the immediate, tangible benefits of their solutions in addressing specific threats. This can be highly persuasive. For instance, during periods of heightened geopolitical tension or following a significant cyberattack, vendors are quick to offer their expertise and technologies as immediate remedies. This urgency can override more deliberate, long-term policy considerations, such as fostering open standards, promoting interoperability, or investing in indigenous research and development capabilities that are not tied to foreign commercial interests. The "fear, uncertainty, and doubt" (FUD) often associated with cybersecurity threats is a powerful tool that vendors can leverage to accelerate procurement and influence policy decisions in their favor. It is, once more, Ian Levy's magic amulet, now being sold not to a single buyer but written into the rulebook every buyer must follow.

The challenge for governments and regulatory bodies is to harness the invaluable expertise of the cybersecurity industry without allowing commercial interests to dictate policy. This requires a conscious effort to foster diverse perspectives in policy development, to rigorously scrutinize vendor proposals for inherent biases, and to prioritize the development of open standards and interoperable solutions. It also necessitates robust independent research and testing capabilities to validate vendor claims and to explore alternative approaches to cybersecurity. The goal should be to create an environment where the most effective and resilient security outcomes are achieved, rather than simply facilitating the sale of proprietary products. The influence of vendors is a critical factor to understand in the complex ecosystem of OT cybersecurity, and recognizing its manifestations is the first step towards mitigating its potential downsides and ensuring that policy truly serves the public interest and national security.

Taken together, these mechanisms produce a de facto endorsement of certain technologies. A standard need not name a vendor; its specifications can simply be so closely aligned with a particular product that compliance becomes synonymous with purchase. The same dynamic governs certifications and accreditations: when a mandated certification process is shaped by vendor input or proprietary methodologies, the vendors who helped write the requirements design their products to meet them, gaining a decisive advantage and erecting a barrier to entry for newer, possibly better, competitors. None of this requires overt corruption. It requires only that commercial interests be allowed to shape technical discourse, and that the urgency of the moment, often manufactured by the very firms standing to benefit, crowd out the slower, less profitable work of building genuine resilience. Policies that fixate on procuring discrete tools, malware detection here, a threat-intelligence feed there, at the urging of vendors promoting their latest offerings, yield a fragmented and reactive posture: a costly suite of point solutions that, lacking cohesion and integration, leaves the organization less secure than a deliberately designed system would.

Fear Is the Business Model

Everything described so far in this chapter, the fear-as-a-service marketing, the theater of the hack, the opacity of efficacy, the gilded cage of lock-in, the quiet capture of the rulebook, has until recently been understood as a set of *sales tactics*. Tactics a vendor uses to move a product. But a deeper and more uncomfortable truth has come into focus over the past decade, one this book argues is the real story of the modern cybersecurity market. Fear is no longer merely the tactic. Fear is the asset. The threat narrative itself, the shared, escalating, never-quite-falsifiable conviction that catastrophe is imminent, has been financialized. It is bought, marked up, packaged, and sold, and the returns it generates have become the gravitational center around which the entire industry now orbits.

The intellectual groundwork for this argument was laid more than a decade ago, though its authors could not have foreseen how literally it would come true. In 2011, the economists Jerry Brito and Tate Watkins published as a Mercatus Center working paper at George Mason University and, in the same year, in the Harvard National Security Journal, titled "Loving the Cyber Bomb? The Dangers of Threat Inflation in Cybersecurity Policy."[20] In it they coined a phrase that has only grown more apt with time: the "cyber-industrial complex." Such a complex, they warned, "may serve to not only supply cybersecurity solutions to the federal government, but to drum up demand for them as well."[21] And in a companion essay that same year, they put their finger on the structural conflict of interest at its heart: "Washington," they wrote, "is filled with people who have a vested interest in conflating and inflating the threats to our digital security."[22]

Brito and Watkins were describing a specific machine: the *government-spending* version of the cyber-industrial complex, in which defense contractors and agencies talk one another into ever-larger federal appropriations. That machine is real, and this book has already traced its workings. But in the years since 2011, a second and in many ways more powerful engine has been bolted onto the first, one the original authors could not fully have anticipated: the *private-capital* version. The cyber-industrial complex of the 2020s is driven not only by federal budgets but by private equity and venture capital, by firms whose entire business is to buy the threat narrative cheap, ride its appreciation, and sell it dear. To understand how fear became an asset class, follow the money.

Begin with the simplest demonstration: that the threat narrative is literally an appreciating asset, an item whose price rises over time on the strength of nothing more than sustained anxiety. In 2019, the private-equity firm Thoma Bravo acquired the web-and-data-security company Imperva for approximately $2.1 billion. In December 2023, it agreed to sell Imperva to the French aerospace and defense group Thales (the deal closed in 2024) for approximately $3.6 billion.[23] The underlying product did not transmute into something $1.5 billion better in four years. What appreciated was the franchise, and the franchise is, at bottom, a claim on a future of persistent threat. The pattern repeats at greater magnitude with Armis, an asset-visibility and OT-security firm.

The venture-and-growth investor Insight Partners bought into Armis at a valuation of roughly $1.1 billion in 2020, saw that stake marked to roughly $3.4 billion in 2021, and, in December 2025, agreed to sell the company to ServiceNow for $7.75 billion, a deal ServiceNow completed on April 20, 2026.[24] Roughly seven times the entry valuation in five years, on the same essential thesis: that the world will remain frightened enough of industrial cyber threats to keep paying. The threat narrative is the underlying security in the financial sense of that word, and its chart goes up and to the right.

If fear is an appreciating asset, then the natural next move for sophisticated capital is to accumulate it at scale, and that is exactly what has happened. Thoma Bravo, a private-equity firm managing approximately $181 billion in assets as of June 30, 2025, has spent the better part of a decade assembling a cybersecurity empire with the methodical patience of a conglomerate-builder.[25] In 2021 it acquired the email-security company Proofpoint for $12.3 billion.[26] In a single year, 2022, it acquired roughly twelve billion dollars' worth of identity-security companies, buying SailPoint for $6.9 billion, Ping Identity for $2.8 billion, and ForgeRock for $2.3 billion, the last of which it subsequently merged into Ping.[27] It had earlier agreed in 2019 to take the endpoint-security firm Sophos private for approximately $3.9 billion, a deal that closed in 2020, and in April 2024 it agreed to acquire the AI-security company Darktrace at an enterprise value of approximately $5.3 billion.[28] Pause on the identity-security spree in particular. A single private-equity firm consolidated a major share of the identity-security category in the span of twelve months, because the recurring-revenue mathematics of selling subscription protection against an ever-present threat simply work. You do not assemble a category that quickly unless the cash flows beneath it are extraordinarily reliable. And what makes them reliable is the durability of the fear.

Nowhere is the financialization of fear more on the nose than in the firm of Crosspoint Capital. Crosspoint is a private-equity outfit founded by Greg Clark, the former chief executive of Symantec — the company whose Norton consumer antivirus brand did as much as any to build the consumer anxiety about computer viruses that defined the personal-computing era. One of Crosspoint's partners is Hugh Thompson, who also chairs the program for the RSA Conference, the flagship annual gathering of the entire cybersecurity industry and the single most important marketing event in the field. Crosspoint closed its debut fund at $1.3 billion in 2021. Consider what that arrangement actually means. The people who helped manufacture the modern consumer-fear industry now run a fund that financializes it, and one of them simultaneously presides over the calendar's marquee event, the stage on which the year's threats are unveiled and the year's solutions are sold. The supply of fear, the capital that profits from fear, and the platform that broadcasts fear have, in this instance, the same address book. None of this requires a conspiracy. It requires only that everyone follow their incentives, and the incentives all point the same direction: toward a world that stays afraid.

Step back and the flows of capital tell the macro version of the same story. Venture-capital investment into the cybersecurity sector ran to roughly $22.8 billion in 2021, fell to $15.3 billion in 2022, dropped again to $8.1 billion in 2023, and recovered to roughly $11.6 billion in 2024.[29] Billions of dollars of patient, return-seeking capital flow into the sector every single year, and every dollar of it carries an implicit demand: that the threat be large and that it grow. That demand is made explicit in the way the industry talks about its own size. Industry analysts at Gartner projected in 2025 that the global cybersecurity market will reach roughly $322 billion by 2029.[30] More striking still, the consulting firm McKinsey estimated in 2022 a potential total addressable market on the order of $1.5 to $2 trillion, roughly ten times the size of today's market.[31] Read that figure the way an investor reads it. A ten-times-larger market is the investment thesis. It is the number in the pitch deck that justifies every acquisition, every markup, every fund. And here is the part that the thesis cannot escape: that market only grows to its projected size if the fear grows with it. The total addressable market for protection is, definitionally, a function of perceived threat. To bet on the TAM is to bet on sustained, expanding anxiety. The capital is not a neutral bystander to the fear machine. It is the flywheel's heaviest mass.

Bruce Schneier's "security theater" named the demand side of this system: the human tendency to pay for the feeling of safety. Brito and Watkins named the supply side as it existed in 2011: a Washington ecosystem incentivized to inflate threats and harvest federal budgets. What has happened since is that the two have been welded into a single, self-reinforcing machine and handed an accelerant. The flywheel turns like this. Fear generates a narrative of imminent catastrophe. The narrative justifies regulation, the SEC's four-day breach-disclosure clock, the European Union's NIS2 expansion, and the regulation converts the narrative into legally mandated spending. Mandated spending becomes the predictable, recurring revenue that private equity and venture capital prize above all else. That recurring revenue is then capitalized into soaring valuations, Imperva's $1.5-billion appreciation, Armis's sevenfold rise, Thoma Bravo's tens of billions in roll-up, and those returns reward and attract still more capital. And capital, having staked its returns on the size of the threat, acquires a permanent, structural interest in keeping the threat narrative elevated. The output of the flywheel feeds its input. Fear funds the machine, and the machine manufactures fear.

None of this is a victimless abstraction confined to the cap tables of private-equity funds. Someone pays for the markup. When a roll-up acquires a security vendor, the thesis that justifies the purchase price is almost never "we will make this product so effective that customers eventually need less of it." It is the opposite: raise prices, convert onetime licenses into perpetual subscriptions, bundle adjacent products to deepen the lock-in described above, and harvest the compliance-driven recurring revenue that regulation has rendered non-optional. The financial logic rewards a customer who stays subscribed and stays worried, not one who becomes secure enough to walk away. A water utility passes those costs to its ratepayers; a hospital to its patients; a manufacturer to its consumers; and the federal government, the single largest cybersecurity buyer on earth, to its taxpayers. The fear is capitalized on Wall Street and amortized on Main Street.

And because the asset is made of sentiment rather than steel, it is volatile, and when it falls it takes real things with it. IronNet did not destroy a billion dollars of notional value in the abstract: it laid off its engineers, stranded its customers mid-contract, and wiped out the retail investors who had been told they were buying a piece of the national defense. Capital and talent that might have gone toward the unglamorous work of hardening real systems went instead toward a story, and when the story failed the audit, the hardening was never done. This is the hidden tax of a market organized around fear — not merely that buyers overpay, but that the most lucrative thing an engineer or an investor can build comes to be the *appearance* of protection rather than the fact of it.

This is the deepest sense in which the cybersecurity industry profits from fear, deeper than any single misleading demonstration or overstated threat report. It is not that individual salespeople exaggerate, though some do. It is that the financial architecture of the entire sector, the private-equity roll-ups, the venture markups, the trillion-dollar addressable-market projections, now rests on the foundational assumption that the world will remain afraid, and afraid at an accelerating rate. The general who warned of "the greatest transfer of wealth in history" built a company on that warning and saw its share price collapse to roughly twenty-six cents before delisting in 2023. The man who built the consumer-virus scare now runs a fund that financializes it. The equipment makers who sold the vulnerable hardware now own the firms that monitor it for threats. And a sitting government defense official looked at the whole apparatus and called it medieval witchcraft, the selling of a magic amulet to people told they are too stupid to protect themselves. The amulet, it turns out, has a price-to-earnings ratio. Recognizing that, seeing fear not merely as a marketing mood but as the literal asset on which a multi-hundred-billion-dollar industry has been capitalized, is the necessary first step toward asking the only question that matters: not how frightened we have been told to be, but how frightened the evidence actually warrants.

Chapter 5

POLICY FRAMEWORKS BUILT ON MYTH: A CRITICAL EXAMINATION

The foundational documents that guide a nation's approach to cybersecurity, particularly those pertaining to its critical infrastructure and national security, often present a powerful and compelling narrative of impending doom. These strategies, while ostensibly designed to protect vital systems, frequently lean heavily on the specter of catastrophic cyberattacks. The critical question, however, is whether this narrative is grounded in a rigorous, empirical assessment of current and emerging threats to Operational Technology (OT) environments, or if it has become a self-perpetuating "fear machine," amplified by various interests and driven by a focus on hypothetical, large-scale disruptions rather than the fundamental governance and operational challenges that plague OT security. In the OT context, the distinction that matters for this chapter is narrow but decisive: these are systems whose failures register not as lost data but as physical consequences, and policy written for the former rarely fits the latter.

When examining the National Security Strategy (NSS) and its related directives concerning OT cybersecurity, a recurring theme emerges: the emphasis on sophisticated, state-sponsored actors capable of orchestrating widespread damage to industrial control systems (ICS) and other critical infrastructure. These documents often paint a picture of adversaries meticulously probing defenses, possessing advanced capabilities to penetrate air-gapped networks, and standing ready to unleash devastating cyber-physical attacks. The potential consequences are invariably described in stark terms: widespread power outages, disruption of water supplies, collapse of transportation networks, and cascading failures across multiple sectors, leading to significant economic and social upheaval. This framing is not without its intended effect; it serves to galvanize political will, justify increased budgets, and mobilize resources towards what is presented as an existential threat. These are not private fears confined to classified briefings; they are public documents, and the responsible thing for any critical reader to do is to read what they actually say. [1]

The Disconnect with On-the-Ground Reality

However, a critical examination reveals a disconnect between the articulated threat and the documented reality of OT security incidents. While sophisticated attacks do occur, and the potential for harm is undeniable, the historical record of successful, large-scale cyber-physical attacks directly attributable to state-sponsored actors targeting critical OT infrastructure remains remarkably thin. High-profile incidents, while alarming and instructive, often reveal vulnerabilities stemming from basic cybersecurity hygiene failures, misconfigurations, or the exploitation of widely known vulnerabilities, rather than the deployment of highly novel, nation-state-level capabilities against well-defended systems. The Stuxnet worm, often cited as a prime example of sophisticated OT cyber warfare, targeted specific industrial processes and required a highly tailored approach, and even then, its full impact and attribution remain subjects of academic and intelligence debate. More recent incidents, such as the Colonial Pipeline ransomware attack, while severely impactful, were primarily driven by financial motives and exploited common IT vulnerabilities that had, regrettably, extended into OT environments.

This disparity raises crucial questions about the underlying basis of national security strategies concerning OT. Are these strategies reflecting the most probable risks to OT, or are they disproportionately influenced by worst-case scenarios extrapolated from IT threat landscapes? The very nature of OT, its legacy systems, specialized protocols, and unique operational imperatives, presents a different risk profile than traditional IT environments. The focus on espionage, data exfiltration, or denial-of-service attacks, which are prevalent in IT, may not always translate directly to the OT domain, where the primary concern is often the safety and continuity of physical processes. Yet, national security strategies frequently adopt a generalized cybersecurity framework that, while applicable in principle, may not adequately address the nuanced and specific vulnerabilities inherent in OT.

Anatomy of the Fear Machine

The perpetuation of this "fear machine" can be attributed to several interlocking factors. Firstly, there is an inherent tendency for national security discourse to highlight the most alarming potential threats. This is a natural consequence of seeking to demonstrate the necessity of intelligence gathering, defense spending, and proactive security measures. Policymakers and intelligence agencies are often incentivized to report on the most sophisticated and dangerous adversary capabilities, even if the probability of their deployment against specific targets is low. This emphasis on the extreme, while serving certain purposes, can inadvertently overshadow the more mundane, yet equally dangerous, reality of insecure OT systems.

This is not a vague tendency to be asserted in the abstract; it has a documented, on-the-record history, delivered by named officials from the most authoritative podiums in American government. The archetype was set on October 11, 2012, when Secretary of Defense Leon Panetta, addressing the Business Executives for National Security aboard the decommissioned aircraft carrier Intrepid, warned of a coming "cyber Pearl Harbor." An "aggressor nation or extremist group," he cautioned, could "derail passenger trains, or even more dangerous, derail trains loaded with lethal chemicals. They could contaminate the water supply in major cities, or shut down the power grid across large parts of the country." [2] The imagery was precise, the consequences apocalyptic, and the venue calculated for maximum effect. In the more than a decade since, none of it has occurred. No chemical train has been derailed by an adversary's keystroke, no major U.S. city's water has been poisoned through its control systems, and no American regional grid has gone dark at the hand of a hostile state. The scenario remains exactly what it was when Panetta described it, a vivid hypothetical.

Two years later, the pattern repeated at an even higher level of specificity. On November 20, 2014, Admiral Mike Rogers, then director of the National Security Agency and commander of U.S. Cyber Command, told the House Intelligence Committee that "it is only a matter of the when, not the if, that we are going to see something traumatic" against American infrastructure, framing a major attack as effectively inevitable within roughly the coming decade. [3] What is instructive here is not merely the confidence of the prediction but its self-imposed deadline. That horizon has now passed. The decade Rogers gestured toward has elapsed without the "something traumatic" he forecast, which ought to give any policymaker pause before treating the next such warning as actionable certainty rather than institutional habit.

The most recent and most rhetorically elaborate entry in this lineage came from Jen Easterly, director of the Cybersecurity and Infrastructure Security Agency (CISA), in testimony before the House Select Committee on the Chinese Communist Party on January 31, 2024. "This is truly an Everything Everywhere, All at Once scenario," Easterly told the committee, borrowing the title of an Academy Award–winning film to dramatize the stakes. "Imagine not one pipeline, but many pipelines disrupted," she continued. "People start getting sick from polluted water. Trains get derailed." [4] The catalogue is, almost word for word, the same catalogue Panetta recited twelve years earlier, the same trains, the same water, the same cascading collapse, now repackaged through a pop-culture reference for a new news cycle.

The tell, as it so often is, lies in the single concrete example Easterly could actually point to. That example was Colonial Pipeline, and the testimony was delivered while CISA was actively appealing to Congress for funding. But Colonial, as established earlier in this book, was criminal ransomware that struck the company's billing systems; it was Colonial's own leadership that elected to shut the pipeline down out of caution, not an adversary seizing the controls. The one real-world data point on offer was not a controls takeover at all. The catastrophic version, the one that would actually justify the rhetoric, remained where it always remains, in the conditional mood: "imagine." When the single concrete instance an official can cite turns out, on inspection, to be a financially motivated IT intrusion that triggered a voluntary operational shutdown, and the genuinely catastrophic claims are all prefaced by "imagine," the structure of the argument has been laid bare.

It would be unfair, and analytically lazy, to suggest that this hyperbolic register is universal among senior officials, or that no one in a position of authority has ever pushed back on it. Some have, pointedly. In a Statement for the Record before the House Intelligence Committee on September 10, 2015, Director of National Intelligence James Clapper is reported to have formally rejected the "cyber Armageddon" and "cyber Pearl Harbor" framing, cautioning instead that the more realistic near-term threat was an ongoing series of lower-level, costly intrusions rather than a single civilization-ending blow. [5] That the nation's most senior intelligence official found it necessary to talk his own colleagues and the public down from the "Pearl Harbor" metaphor is itself evidence of how deeply embedded the metaphor had become, and how little it corresponded to the assessed reality even within the intelligence community that supposedly underwrote it.

The longevity of the metaphor deserves its own scrutiny, because it reveals the machine's most durable feature: its unfalsifiability. The phrase "Electronic Pearl Harbor" did not originate with Panetta or any sitting official. It traces to 1991 and the security writer Winn Schwartau, who deployed it before the commercial internet existed in any recognizable form. [6] Across the three decades since, the prophecy has been sustained not by fulfillment but by a recurring escape-hatch: when the predicted cataclysm fails to arrive on schedule, its proponents simply assert that it is already underway, only invisibly. Richard Clarke, the former counterterrorism and cybersecurity adviser, offered a representative version of this maneuver in 2012 when he spoke of "the little Pearl Harbors that are happening every day." [7] The rhetorical move is elegant and corrosive in equal measure: a prediction that can be neither confirmed by a visible catastrophe nor refuted by its absence, because absence itself is reinterpreted as a catastrophe in progress. A claim constructed this way is, by design, immune to the historical record.

If this sounds like an isolated reading, the academic literature corroborates it directly. In their 2019 First Monday study, Sean Lawson and Michael Middleton conducted a content analysis of roughly twenty-five years (1991–2016) of news coverage invoking the "cyber Pearl Harbor" trope. Their finding cuts against any assumption that journalism grew more skeptical as the predicted catastrophes repeatedly failed to materialize. The opposite occurred: coverage became more credulous over time, with on the order of seventy-seven percent of the analyzed coverage treating the metaphor neutrally and only about three percent adopting a skeptical posture, even as the count of actual civilization-scale attacks stayed at zero. [8] A narrative that grows more believed precisely as it accumulates a longer record of nonoccurrence is not responding to evidence. It is the signature of a fear machine operating independently of the facts it claims to describe.

Secondly, the cybersecurity industry plays a significant role in shaping the threat narrative. Vendors of security solutions, eager to market their products and services, often leverage the perceived threat landscape to create demand. Their marketing materials, white papers, and presentations frequently highlight the most sophisticated attack vectors and the most catastrophic potential outcomes. When these narratives are then echoed in policy discussions and national security documents, a feedback loop is established, where the industry's depiction of threats influences policy, which in turn creates a market for the industry's solutions. This is not a conspiracy, but a natural alignment of commercial interests with the prevailing security discourse. A vendor specializing in

OT anomaly detection, for instance, will naturally emphasize the likelihood of stealthy, sophisticated intrusions that their technology is designed to detect.

Thirdly, the lack of transparency and independent assessment within many OT sectors makes it difficult to obtain an objective view of the threat landscape. Critical infrastructure operators, often driven by regulatory compliance and a desire to avoid negative publicity, may be hesitant to publicly disclose the full extent of their vulnerabilities or past incidents. This information asymmetry allows generalized threat narratives, often amplified by vendors and security researchers, to fill the void. Without robust, independent data on the actual frequency, impact, and root causes of OT security incidents, policymakers are left to rely on the most vocal and well-resourced sources, which often tend to be those with a commercial stake in the cybersecurity market.

The Consequences of a Fear-Based Strategy

A national security strategy focused on hypothetical attacks at the expense of foundational issues leads to significant consequences, primarily a misallocation of resources. Investments may be directed towards advanced threat detection and response technologies designed to counter sophisticated state actors, while neglecting fundamental security practices such as asset inventory, vulnerability management, secure configuration, access control, and robust incident response planning. These foundational elements are often the weakest links in OT security and are far more likely to be exploited by a wider range of adversaries, including less sophisticated attackers, insiders, or even accidental misconfigurations.

Secondly, such strategies can create a false sense of security. If the narrative focuses on advanced nation-state threats, organizations might believe they are adequately protected if they have implemented some of the more sophisticated, often vendor-provided, solutions. However, they may remain highly vulnerable to common, well-understood threats that are not as dramatic or as profitable to address through cutting-edge technology. The emphasis shifts from achieving baseline resilience to defending against publicized, though less probable, scenarios.

Thirdly, an overreliance on hypothetical catastrophic events can stifle the development of practical, effective security measures tailored to the specific needs and constraints of OT environments. Many OT systems are designed for longevity, reliability, and safety, often with limited patching capabilities, proprietary protocols, and unique operational constraints. A "one-size-fits-all" approach to cybersecurity, derived from

IT best practices and amplified by the urgency of national security, may not be appropriate or even feasible for these environments. The focus on preparing for an imagined, devastating attack can detract from the incremental, painstaking work of improving the security posture of existing systems and fostering a culture of security awareness among OT personnel.

Moreover, the very definition of "critical infrastructure" and the prioritization of its protection within national security strategies can be influenced by this narrative. Sectors that are most susceptible to the types of large-scale disruptions emphasized in strategic documents, such as energy and utilities, may receive disproportionate attention and funding, while other critical sectors, perhaps less prone to dramatic cyber-physical attacks but still vital for national functioning, might be overlooked. This can lead to an unbalanced approach to national cybersecurity, where resources and attention are concentrated on the most sensational threats, rather than on a comprehensive risk assessment across all essential services.

The policy documents themselves often employ language that, while intended to convey urgency, inadvertently amplifies the fear. Phrases like "unprecedented threats," "imminent dangers," and "existential risks" recur across strategy documents and the public statements that accompany them, and as the testimony catalogued above demonstrates, they are rarely tethered to a specific, observed event. They function instead as mood-setting. When Easterly reaches for "Everything Everywhere, All at Once," when Rogers invokes the inevitability of "something dramatic," when an executive order's preamble warns of threats to the very functioning of the nation, the cumulative effect is a generalized atmosphere of alarm that becomes detached from actionable intelligence about actual threats to OT. This creates an environment where any disruption, however minor or common, can be sensationalized as a potential precursor to a catastrophic event, further fueling the fear machine.

This generalized alarm carries a consequence the fear machine never books, because the bill comes due only later: the steady erosion of its own credibility. A warning system that forecasts catastrophe every year, and is wrong every year, does not keep a population vigilant — it teaches that population to stop listening. The testimony catalogued above spans more than three decades: an "electronic Pearl Harbor" in 1991, a "cyber Pearl Harbor" in 2012, "something dramatic" promised before 2025. The predicted cataclysm has not arrived. Each unfulfilled prophecy is spent credibility drawn from an account no one replenishes, and the cost is borne not by the official who issued the warning — who pays no penalty for crying wolf — but by the next official, who may possess a genuine signal and find a shrinking audience willing to credit it. At the operational level the same dynamic already has a clinical name, *alert fatigue*: control-room staff buried under a ceaseless torrent of "critical" alarms learn, as a matter of cognitive survival, to triage the urgent into the ignorable. A strategy that brands every anomaly a potential Pearl Harbor thus manufactures the very complacency it claims to be fighting. The parable of the boy who cried wolf is not, in the end, a story about the wolf. It is a story about the village's ears.

Consider the recurring emphasis on the "air gap." National security strategies often speak of adversaries overcoming air gaps to compromise critical OT systems. While the concept of an air gap as a perfect security measure is largely a myth in modern interconnected environments, its continued invocation in policy documents suggests an underlying assumption that traditional, perimeter-based security thinking is still relevant, or that breaches of such defenses are the primary modus operandi of sophisticated attackers. This focus on a specific, albeit often breached, defense mechanism can distract from the more pervasive and insidious threats that arise from supply chain compromises, insider threats, or the simple failure to patch known vulnerabilities, which are often far more prevalent and easier to exploit than breaching a hypothetical air gap.

Furthermore, the tendency to view OT cybersecurity solely through the lens of nation-state adversaries can obscure the significant risks posed by criminal enterprises, hacktivists, and even unintentional human error. While nation-state attacks are a serious concern, ransomware gangs, for example, have demonstrated a significant capacity to disrupt critical services, not through sophisticated cyber-physical attacks, but through the effective exploitation of common vulnerabilities and the disruption of essential business and operational processes. The Colonial Pipeline episode is the illustrative case: the most consequential pipeline shutdown in recent American memory was set in motion not by a state-sponsored controls takeover but by financially motivated criminals and a cautious corporate decision. National security strategies that primarily focus on state-level threats may not adequately address the diverse array of actors and motivations that pose a real and present danger to OT infrastructure.

The development of effective policy requires a foundation of accurate, granular, and continuously updated threat intelligence. When that intelligence is filtered through a lens of sensationalism, or when the narrative is disproportionately shaped by commercial interests, the resulting policies risk being misaligned with the actual risks. This is not to say that policymakers are intentionally misleading, but rather that the complex ecosystem of threat assessment, policy formulation, and industry influence can inadvertently lead to strategies that prioritize the perception of security over the achievement of actual resilience. The Lawson and Middleton finding bears repeating in this connection: a discourse that grows more credulous as its predictions go unfulfilled is structurally incapable of self-correction, and policy built atop such a discourse inherits the same defect.

Shifting national security strategy from sensationalism to a grounded assessment of probable risks requires a commitment to several principles:

1. **Data-Driven Risk Assessment:** Prioritizing the collection and analysis of real-world data on OT security incidents. This includes not only successful breaches but also near misses, attempted attacks, and common vulnerabilities exploited. Independent research and government agencies need to collaborate to build a comprehensive, anonymized database of OT security events.

2. **Focus on Foundational Security:** Emphasizing the importance of basic cybersecurity hygiene and robust governance principles as the bedrock of OT security. This includes asset management, configuration management, access control, patching, network segmentation, and comprehensive incident response planning. These are not glamorous, but they are the most effective defenses against a wide range of threats.

3. **Tailored OT Security Approaches:** Recognizing that OT environments have unique characteristics and risk profiles that require specialized security solutions and strategies. Policies should encourage the development and adoption of frameworks and technologies that are specifically designed for OT, rather than simply adapting IT-centric approaches.

4. **Transparency and Information Sharing:** Fostering an environment where critical infrastructure operators feel safe to share information about vulnerabilities and incidents without fear of undue penalty. Public-private partnerships focused on threat intelligence sharing and collaborative defense are crucial.

5. **Independent Verification and Validation:** Developing robust mechanisms for independently verifying the effectiveness of security claims made by vendors and the efficacy of implemented security measures, rather than relying solely on vendor assurances or theoretical threat models.

6. **Addressing the "Human Factor":** Recognizing that operational personnel, engineers, and IT staff are often the first line of defense, and that training, awareness, and clear operational procedures are critical components of security, often overlooked in technology-centric strategies.

Without such a recalibration, national security strategies risk becoming a self-fulfilling prophecy, where the constant emphasis on extreme threats leads to a neglect of the fundamental vulnerabilities that are more likely to be exploited, thereby perpetuating a cycle of reactive security and increasing the nation's actual risk. The "fear machine," once set in motion, can be difficult to dismantle, but a critical examination of the underlying assumptions and motivations is essential to ensure that national security policy truly serves to enhance, rather than merely sensationalize, the protection of vital OT infrastructure. The focus must evolve from the abstract possibility of a nationwide blackout orchestrated by a rival superpower, to the concrete reality of ensuring that the power grid, the water systems, and the transportation networks are resilient against the full spectrum of threats, from the sophisticated to the mundane, and from external adversaries to internal missteps.

The Ambiguous Role of Information Sharing Centers (ISACs)

The genesis of the ISAC model lies in the recognition that no single entity, whether governmental or private, possesses a complete view of the threat landscape or the resources to defend against it alone. The model posits that by creating sector-specific communities where organizations can voluntarily share sensitive information about cyber threats, vulnerabilities, and incidents, a collective defense can be forged. This shared intelligence, proponents argue, allows participants to proactively adapt their defenses, anticipate emerging attack vectors, and respond more effectively when incidents occur. For OT environments, where the consequences of compromise can extend beyond data loss to physical disruption and safety hazards, the theoretical promise of ISACs is particularly potent. Imagine a scenario where an electric utility, a water treatment plant, and a transportation network operator, all members of the same energy or infrastructure ISAC, receive early warnings about a novel exploit targeting a common industrial control system component. Armed with this timely intelligence, each organization could swiftly implement patches, reconfigure firewalls, or enhance monitoring, potentially averting a significant disruption.

However, the practical realization of this ideal in the OT domain has often fallen short. One of the primary challenges is the inherent reluctance of many critical infrastructure operators to share detailed information about their security posture, vulnerabilities, or actual incidents. This reticence stems from a confluence of factors, including regulatory pressures, competitive sensitivities, the fear of reputational damage, and the potential for legal liability. Disclosing a significant security lapse, even to a trusted ISAC, could invite scrutiny from regulators, attract negative media attention, or expose the organization to lawsuits from stakeholders or customers. Consequently, the information that is shared might be sanitized, anonymized to the point of reduced actionable insight, or focused on generalized threat indicators rather than specific, exploitable details. This can transform the ISAC from a dynamic intelligence hub into a repository of rather generic cybersecurity advisories, which may not adequately prepare organizations for the nuances of OT-specific threats.

Furthermore, the governance and operational structure of many ISACs can inadvertently create an environment ripe for what some critics term "threat inflation." The mechanism is best understood not as malice but as a funding-incentive structure. Many ISACs are sustained by a combination of member organizations' dues and the sponsorship of cybersecurity vendors, and that revenue model quietly shapes editorial priorities. Vendors are essential partners in providing solutions and expertise, and their participation in ISACs, often as service providers, sponsors, or advisors, is in many respects valuable. But the arrangement carries a structural bias. The ISAC needs engaging content to justify its dues and retain its sponsors; the vendor needs a market for its products; and the path of least resistance for both is an overemphasis on sophisticated, dramatic, and perhaps less probable threats. White papers, webinars, and threat reports produced or influenced by sponsoring vendors tend to highlight the most advanced attack methodologies, the very ones their cutting-edge solutions are built to detect and mitigate. That emphasis then filters into the ISAC's communications to its members, reinforcing the "fear machine" described earlier. Organizations, inundated with alerts about state-sponsored advanced persistent threats targeting OT systems with zero-day exploits, might divert resources and attention away from more fundamental, yet equally critical, security practices like robust access control, patch management for legacy systems, or comprehensive personnel training, areas where the commercial return for a sponsoring vendor is less immediate or headline-grabbing. The point is not that any particular ISAC has been captured, but that a funding model built on dues and vendor sponsorship will, absent deliberate guardrails, tilt toward the threats that are most marketable rather than the threats that are most probable.

The effectiveness of ISACs is also contingent upon the technical expertise and operational context of their member organizations. OT environments are incredibly diverse, ranging from highly specialized legacy systems in nuclear power plants to widely deployed SCADA systems in water utilities, each with unique protocols, hardware, and operational constraints. A "one-size-fits-all" approach to threat intelligence and mitigation advice, which can be a natural consequence of broad ISAC membership, may not be suitable for all sectors or even for all organizations within a single sector. What constitutes a critical vulnerability for a modern, networked smart grid operator might be an irrelevant concern for a facility still relying on air-gapped, proprietary control systems. For ISACs to truly serve the OT community, they must be capable of providing granular, sector-specific, and even contextually relevant intelligence that accounts for these diverse operational realities. This requires a deep understanding of OT not just from a theoretical cybersecurity perspective, but from an industrial engineering and operational standpoint, a level of expertise that can be challenging to maintain across a broad-based organization.

The Colonial Pipeline ransomware attack, while primarily an IT-centric breach that impacted OT operations, serves as a salient example of how shared intelligence and collaborative defense can be both facilitated and complicated by ISAC structures. While the energy ISACs, such as the E-ISAC, undoubtedly played a role in disseminating information and coordinating responses after the incident, the initial compromise itself highlighted a failure in basic IT security hygiene that had extended its tendrils into operational systems. The question then arises: to what extent did the shared intelligence within the ISAC ecosystem adequately address the fundamental vulnerabilities that enabled the ransomware attack, as opposed to focusing on more sophisticated, state-level threats? If the ISACs were indeed flagging common ransomware tactics or advising on better network segmentation between IT and OT, were these warnings sufficiently heeded by all members, or were they overshadowed by more alarming, albeit less probable, threat narratives? The incident underscored that while advanced threat intelligence is valuable, the remediation of fundamental security weaknesses often offers the most significant uplift in resilience. It is worth noting that this is the same incident a sitting CISA director would later hold up, in 2024, as the concrete anchor for a catalogue of hypothetical catastrophes, a telling reminder of how a fundamentally mundane hygiene failure can be retrofitted into the fear narrative.

Moreover, the financial sustainability of ISACs can also influence their operational focus. As noted, many ISACs rely on membership fees, grants, and sometimes vendor sponsorship. This funding model can create a subtle pressure to demonstrate value and relevance to members and sponsors, potentially leading to the prioritization of reporting on high-profile, sophisticated threats that generate significant interest and justify the ongoing investment in security solutions. The continuous, painstaking work of improving asset visibility, conducting regular vulnerability assessments on legacy equipment, or ensuring robust physical security for control system components, while crucial for OT security, might not always be as compelling for inclusion in a quarterly threat brief as the latest bulletin on a nation-state's novel exploitation technique. This can lead to a skewed perception of risk, where the more publicized and technically complex threats capture disproportionate attention, while the everyday risks, often exploited by less sophisticated actors or through sheer negligence, are relatively under-addressed.

The concept of "analysis" within Information Sharing and Analysis Centers also warrants critical examination. Are these centers truly performing deep, independent analysis of raw threat data, synthesizing it into actionable intelligence tailored for OT operators? Or are they primarily acting as conduits for information distributed by government agencies (like CISA or intelligence communities) or aggregated and repackaged by cybersecurity vendors? The level of sophistication and independence in analysis directly impacts the quality and utility of the intelligence provided. If the analysis is superficial or heavily influenced by external commercial pressures, the ISAC's contribution to genuine risk reduction may be marginal. True analysis would involve not just identifying a threat, but also assessing its relevance and applicability to specific OT environments, understanding its likely impact on physical processes, and providing practical, prioritized mitigation strategies. This level of deep analysis requires significant domain expertise and resources, which can be difficult for some ISACs to consistently muster.

The narrative surrounding ISACs, much like the broader cybersecurity policy landscape, is often one of intended purpose versus achieved outcome. The intention is clear: to create a collaborative, informed, and resilient ecosystem for critical infrastructure protection. The reality, however, is more nuanced and potentially problematic. While ISACs have undoubtedly facilitated some level of information exchange and raised awareness among their members, their role in directly improving the security posture of OT environments against the most probable and impactful threats requires ongoing scrutiny. They may be inadvertently contributing to a focus on advanced threats at the expense of foundational security, becoming platforms for vendor-driven narratives, or struggling to provide the granular, context-specific intelligence that OT operators truly need.

A more effective model for ISACs in the OT space would necessitate a significant shift in focus. Firstly, there needs to be a stronger emphasis on data collection and analysis that reflects the actual incident landscape within OT sectors, rather than relying solely on externally generated threat intelligence. This means encouraging, and perhaps incentivizing, members to share more granular, anonymized data about compromises, near misses, and vulnerabilities, and empowering the ISAC to conduct rigorous, independent analysis of this data. Secondly, ISACs should actively work to deprioritize the sensational and focus on the actionable. This involves developing robust frameworks for assessing the relevance and impact of threats specifically on OT processes, and providing clear, prioritized guidance on mitigation, starting with the fundamentals. Thirdly, the relationship with vendors needs to be carefully managed to ensure that their participation serves the collective good without unduly influencing the threat narrative or dictating the focus of security efforts. Because the bias is structural rather than personal, the remedy must be structural too: clearer guidelines on vendor and sponsor contributions, firewalls between sponsorship revenue and editorial content, a greater emphasis on independent verification of vendor claims, and a standing commitment to prioritizing foundational security practices over the promotion of niche, high-end solutions.

Ultimately, the question of whether ISACs contribute to the "fear machine" or genuine risk mitigation hinges on their ability to transcend their often-idealized conceptual framework and deliver concrete, evidence-based improvements in OT security. Without a transparent, data-driven approach that prioritizes the unique challenges and operational realities of industrial control systems, ISACs risk becoming well-intentioned but ultimately ineffective intermediaries, perpetuating a cycle of reactive security and amplified anxieties rather than fostering a truly resilient operational technology landscape. Their success must be measured not by the sophistication of the threats they report on, but by the tangible reduction in successful attacks and the increased resilience of the critical infrastructure they are tasked with protecting.

Compliance vs. Security: The Box-Ticking Dilemma

Many organizations operating critical infrastructure find themselves in a perpetual state of trying to satisfy a mosaic of regulatory requirements. These mandates, originating from diverse sources such as cybersecurity frameworks like NIST, sector-specific regulations (e.g., NERC CIP for the electricity sector, TSA directives for pipelines and rail), and broader data protection laws, establish a baseline of expected security measures. Compliance, in this context, often translates to demonstrable adherence to documented policies, procedures, and technical controls. This focus on tangible, auditable evidence, the presence of an access control list, the execution of a security awareness training session, the implementation of a specific type of firewall, becomes the primary objective. The goal becomes passing an audit and avoiding penalties, not enhancing resilience against novel threats.

This prioritization of compliance over genuine security can manifest in several detrimental ways. One significant issue is the inherent inflexibility of many regulatory frameworks. Laws and standards, by their very nature, tend to lag behind the rapidly evolving threat landscape. By the time a new vulnerability is identified, a threat actor's technique is analyzed, and a regulatory amendment is proposed, debated, and implemented, the cyber domain has already moved several steps forward. Organizations that diligently focus on adhering to outdated or incomplete regulations might find themselves protected against yesterday's threats but left vulnerable to today's emerging challenges. The attack on the Colonial Pipeline, for instance, while involving ransomware that primarily targeted IT systems, ultimately disrupted OT operations. The incident highlighted how even adherence to established IT security policies, which are often more mature than OT-specific ones, might not prevent cascade failures when IT and OT are not adequately segmented and protected. If the regulatory focus was purely on the IT perimeter, the risk to OT could be underestimated and inadequately addressed.

Furthermore, the prescriptive nature of many regulations can stifle innovation and the adoption of more effective, albeit less standardized, security solutions. When organizations are primarily concerned with meeting checklist requirements, they may be disinclined to invest in advanced technologies or methodologies that fall outside the defined scope of compliance, even if those solutions offer superior protection. The "good enough" mentality, driven by the need to achieve compliance with minimal expenditure, can become pervasive. This can lead to a situation where an organization possesses a compliant security posture on paper but remains fundamentally insecure in practice. The cost of implementing and maintaining compliance, especially for legacy OT systems that are difficult to update or integrate with modern security tools, can also be substantial. This financial burden can divert resources away from proactive security measures that might offer a more significant return on investment in terms of risk reduction.

The concept of "defense-in-depth" is a cornerstone of robust cybersecurity, advocating for multiple layers of security controls. However, when compliance is the driving force, defense-in-depth can become a shallow exercise. Instead of strategically layering controls based on a thorough risk assessment of the specific OT environment, organizations might simply implement all the controls mandated by various regulations, regardless of their actual efficacy or relevance. This can result in a collection of disparate security tools and processes that lack integration and strategic coherence. A common scenario involves deploying standard enterprise security solutions into OT environments without proper adaptation, leading to performance issues, operational disruptions, or a false sense of security. For example, a stringent intrusion detection system configured for IT networks might overwhelm a low-bandwidth, real-time OT network, leading to the disabling of critical security alerts to maintain operational uptime. Compliance might mandate the presence of the IDS, but genuine security would require its careful tuning and integration with OT-specific protocols and performance considerations.

The audit process itself can inadvertently reinforce this compliance-centric approach. Auditors, often tasked with verifying adherence to specific regulatory clauses, are not always equipped with the deep OT domain knowledge required to assess the true security posture of industrial control systems. Their focus is on evidence: "Is this document present? Was this training conducted? Is this configuration setting in place?" They may not possess the expertise to question whether these measures are truly effective in mitigating the unique risks associated with OT, such as the physical consequences of cyberattacks, the long lifecycles of industrial equipment, or the operational constraints that limit patching and updating. This can lead to a scenario where organizations become adept at "auditing for compliance" rather than "auditing for security." They learn to present information in a way that satisfies auditors, even if the underlying security practices are weak.

Moreover, the complexity of the regulatory landscape itself can become a barrier to genuine security. Organizations operating across multiple sectors or jurisdictions may find themselves subject to conflicting or overlapping regulations. Navigating these intricate requirements demands significant resources, often diverting attention and budget away from direct security enhancements. The energy required to understand, interpret, and implement each regulation can overshadow the more critical task of understanding the specific vulnerabilities of one's own OT environment and the most probable threats targeting it. This administrative burden can be particularly acute for smaller organizations or those with limited cybersecurity expertise.

Consider the development and deployment of cybersecurity policies. While regulations may mandate the existence of such policies, the actual content and implementation are often driven by compliance considerations. Policies might be drafted to mirror the language of regulatory requirements, incorporating specific clauses and compliance metrics. The risk is that these policies become static documents, rarely updated to reflect evolving threats or operational changes, and their enforcement becomes perfunctory. A policy stating that "all OT systems must be patched within 30 days of vulnerability disclosure" is a clear compliance target. However, in practice, many OT systems cannot be patched within such a timeframe due to operational constraints, vendor limitations, or the risk of introducing instability. A genuinely secure approach would involve risk-based exception handling, compensating controls, and a thorough understanding of the operational impact of patching, rather than a rigid adherence to a potentially impossible compliance deadline.

The financial incentives also play a crucial role in this dynamic. Investing in compliance is often seen as a necessary cost of doing business, with clear penalties for non-compliance. Investing in advanced, proactive security measures, on the other hand, can be viewed as a discretionary expenditure with a less predictable return on investment. The "fear machine" discussed previously can also contribute to this, as it often amplifies the perceived threat of sophisticated attacks, driving investment towards specific vendor solutions that promise to protect against these advanced threats, while potentially neglecting fundamental security hygiene that might be mandated by compliance but is less "exciting" for marketing purposes.

The disconnect between regulatory compliance and genuine security is perhaps most starkly illustrated in the realm of legacy OT systems. These systems, often decades old, were not designed with modern cybersecurity threats in mind. They may run on outdated operating systems, use proprietary protocols, and lack basic security features like encryption or robust authentication. Regulations often struggle to address these unique challenges effectively. Mandating specific security controls for systems that cannot support them is an exercise in futility. Compliance efforts might then focus on workarounds, such as implementing network segmentation or compensating controls, which are indeed valuable. However, the underlying vulnerability of the legacy system remains, and the compliance framework may not sufficiently incentivize or demand the costly and complex process of modernization or replacement. This creates a persistent blind spot in the security posture of critical infrastructure.

Another facet of this issue is the attribution of responsibility. Regulatory frameworks often assign security responsibilities to specific roles or departments within an organization. While this helps in defining accountability, it can also lead to siloed thinking. The OT engineers, responsible for keeping the plant running, might view cybersecurity as the IT department's problem, while the IT security team, focused on compliance, may not fully grasp the operational intricacies and risks within the OT environment. Genuine security, however, requires a collaborative, cross-functional approach where IT and OT teams work in tandem, sharing knowledge and jointly developing strategies. When compliance becomes the sole driver, these necessary collaborations can be deprioritized in favor of meeting individual departmental compliance obligations.

The Information Sharing and Analysis Centers (ISACs), while aiming to foster collaboration and intelligence sharing, can also fall victim to this compliance-centric mindset. If the primary driver for participation in an ISAC is to demonstrate due diligence to regulators or auditors, then the focus might shift from actively seeking and applying actionable intelligence to simply collecting reports and meeting attendance requirements. The value derived from the ISAC would then be in its ability to provide documentation of engagement rather than a tangible improvement in security posture. This can lead to a superficial participation where organizations receive threat intelligence but fail to implement the necessary changes because those changes are not mandated by current regulations or are too costly to implement without a clear compliance imperative.

The danger of prioritizing regulatory compliance over genuine security is that it creates a false sense of security. Organizations might believe they are secure because they meet all the regulatory requirements, only to discover during a real-world incident that their defenses are inadequate. This can have catastrophic consequences for critical infrastructure, impacting public safety, economic stability, and national security. The focus should, therefore, shift from merely "checking the boxes" to fostering a culture of proactive risk management, where security is integrated into the operational fabric of the organization, and where decisions are driven by a deep understanding of the specific threats and vulnerabilities, rather than solely by the dictates of external mandates. This requires a fundamental rethinking of how security is approached in OT environments, moving beyond the compliance-driven paradigm to one that embraces continuous improvement, adaptive strategies, and a holistic view of operational resilience.

The challenge lies in bridging this gap. It requires regulators to develop more agile and performance-based frameworks that focus on outcomes rather than specific technical implementations, allowing organizations the flexibility to adopt the most effective solutions for their unique environments. It also necessitates a cultural shift within organizations themselves, where cybersecurity is viewed not as a compliance burden, but as an essential component of operational integrity and business continuity. This shift involves investing in training for both IT and OT personnel, fostering interdepartmental collaboration, and conducting regular, in-depth risk assessments that go beyond the superficial requirements of compliance checklists. Ultimately, true security for OT environments will emerge not from meeting the minimum standards of regulatory compliance, but from a sustained, strategic commitment to understanding and mitigating the real risks that threaten these vital systems.

The Limits of Top-Down Directives

Presidential directives and executive orders are significant levers in shaping national cybersecurity policy. These high-level pronouncements signal national priorities and initiate action across federal agencies. However, their impact on OT security is complex; the intention to bolster defenses can be overshadowed by their role in validating existing paradigms and feeding the threat narrative. Because these are public instruments, their language is available for direct inspection, and the analysis that follows is strengthened wherever the directive's own words are quoted rather than summarized. [9] One of the primary effects of these directives is their capacity to legitimize and amplify the "fear machine." By framing cybersecurity as an existential threat to national security, economic stability, and public safety, executive actions can create a pervasive sense of urgency. This urgency, while understandable given the stakes, can be strategically leveraged to justify increased budgets, new regulatory initiatives, and the adoption of specific security technologies. Presidential directives often highlight dramatic scenarios, detailing potential catastrophic consequences of cyberattacks on critical sectors such as energy, water, and transportation. This emphasis, while intended to spur action, can inadvertently lead to a focus on headline-grabbing threats rather than the more insidious, persistent, and often mundane vulnerabilities that plague OT systems. The narrative often becomes one of facing sophisticated, state-sponsored adversaries or highly organized criminal syndicates, overshadowing the critical need for fundamental cybersecurity hygiene, proper asset management, and robust incident response planning at the foundational level. It is no coincidence that the rhetoric of senior officials testifying in support of these directives—the trains, the water, the grid—maps so closely onto the catastrophic scenarios the directives themselves invoke; the two reinforce each other in a continuous loop.

For instance, a directive might mandate the development of specific cyber threat intelligence-sharing mechanisms between government agencies and critical infrastructure operators. While the intent is to provide operators with timely and actionable information to defend their systems, the practical outcome can sometimes be the creation of more bureaucratic channels for information exchange. The intelligence shared might be high-level, generalized, or focused on the latest exploit rather than providing insights into the specific vulnerabilities of legacy OT systems or the unique operational constraints of industrial environments. This can lead to a situation where organizations are inundated with information but struggle to translate it into concrete security improvements, especially if those improvements require significant investment or operational disruption that is not directly mandated by a compliance checklist derived from the directive. The directive, in this sense, achieves its objective of initiating intelligence sharing, but the effectiveness of that sharing in enhancing OT security remains a separate and often less successful outcome.

Furthermore, presidential directives and executive orders can inadvertently reinforce a compliance-driven approach rather than fostering genuine structural reform. When these directives are translated into agency-specific regulations or guidance documents, they often manifest as lists of required controls, benchmarks to be met, or frameworks to be adopted. While frameworks like the NIST Cybersecurity Framework, often referenced or mandated by executive actions, are valuable tools, their implementation can become a compliance exercise. Operators may focus on "checking the boxes" required by the framework to demonstrate adherence to the spirit of the presidential directive, rather than engaging in a deep, risk-based assessment of their unique OT environment. The directive, in this context, becomes the impetus for creating a compliance program, rather than a catalyst for fundamentally rethinking and re-architecting security for OT.

Consider an executive order that calls for enhanced cybersecurity for the energy sector. This might lead to directives from the Department of Energy or other regulatory bodies that mandate specific security controls for Supervisory Control and Data Acquisition (SCADA) systems and industrial control systems (ICS). While these mandates might include requirements for network segmentation, access controls, and vulnerability management, organizations often implement them in a manner that satisfies the letter of the law without truly enhancing resilience. For example, organizations might implement network segmentation at a high level, but the critical OT networks within those segments may still lack robust internal security measures. Implementers might focus vulnerability management on patching systems that are easily patchable, while they leave critical legacy systems untouched due to operational risks or vendor limitations. This creates significant residual risk that organizations do not adequately address through their compliance regime. The presidential directive, in this scenario, drives activity and spending. Yet, the fundamental security posture of the OT environment often fails to improve in proportion to the perceived threat. Compliance does not equal resilience.

The effectiveness of these policies is diminished by organizational inertia and the complexity of OT environments. Implementing changes mandated by an executive order often requires significant investment in new technologies, extensive training, and changes to deeply ingrained operational procedures. For organizations that have been operating for decades with legacy systems and established practices, these changes are formidable. Presidential directives set direction but often fall short, lacking the granular detail or sustained enforcement needed to overcome systemic challenges. The initial push from an executive order can lead to a flurry of activity, but without continuous government oversight and incentives that go beyond mere penalties for noncompliance, the focus can wane. Organizations then revert to the path of least resistance, often maintaining compliance at the minimum required level. The status quo persists.

Moreover, the focus of presidential directives can sometimes be skewed towards IT-centric security paradigms, which are not always directly applicable or effective in OT environments. Many directives emerge from a context of traditional IT cybersecurity concerns, such as data breaches, intellectual property theft, and network intrusions. While these are valid concerns, OT security involves a different set of risks, including physical consequences of cyberattacks, impacts on public safety, and the unique challenges posed by real-time operations, long system lifecycles, and specialized hardware and software. When directives fail to adequately distinguish between IT and OT security requirements, or when they simply extend IT security mandates to OT without proper adaptation, they can lead to misallocated resources and ineffective security measures.

For instance, an executive order emphasizing the protection of sensitive government data might lead to mandates for stringent encryption and data loss prevention

(DLP) solutions across all connected systems. While crucial for IT, applying these solutions to certain OT systems could be problematic. Encryption might introduce latency or computational overhead that disrupts real-time operations, and DLP solutions might not be compatible with proprietary industrial protocols. In such cases, the directive, while well-intentioned, can inadvertently create a compliance burden that hinders rather than helps OT security, forcing organizations to either compromise operational integrity or find creative, and potentially insecure, ways to comply. The underlying myth that a uniform cybersecurity approach can be applied across all domains, regardless of their fundamental differences, is often perpetuated by such directives.

The role of presidential directives in shaping the "fear machine" is particularly noteworthy. By consistently emphasizing the severity of cyber threats, these directives contribute to a public and private sector perception that the threat landscape is constantly evolving in the most alarming ways. This can lead to a focus on detecting and defending against the most sophisticated and publicized attack vectors, such as advanced persistent threats (APTs) or nation-state-sponsored malware. While these threats are real, they can divert attention and resources from more common, yet equally damaging, threats like ransomware, insider threats, or simple human error. The narrative driven by executive actions often elevates the stakes, making it seem as though the only threats worth addressing are the most catastrophic ones, leading organizations to chase the "biggest boogeyman" rather than focusing on robust foundational security. It is worth recalling that even Director of National Intelligence James Clapper, hardly a dove on cyber matters, is reported to have used his 2015 Statement for the Record to caution Congress against precisely this "cyber Armageddon" framing, favoring instead a sober emphasis on the steady accumulation of lower-grade intrusions. [10] When the official charged with synthesizing the nation's threat intelligence is moved to temper the rhetoric, the gap between the machine's output and the assessed reality is hard to ignore.

This emphasis on the most sophisticated threats can also benefit the cybersecurity industry, fostering a market for advanced, often expensive, solutions that promise protection against these high-level adversaries. Presidential directives can, intentionally or unintentionally, create a demand for such technologies, as organizations seek to demonstrate that they are taking the highest level of threat seriously. This can create a feedback loop where the "fear machine" is fueled by the perceived need for advanced solutions, and the industry benefits from the resulting market demand, all legitimized by the pronouncements of the executive branch. The focus shifts from fundamental risk management to acquiring the latest defense against the most publicized threat, a dynamic that benefits from, rather than fundamentally solves, the challenges of OT security.

Furthermore, the analysis of these directives must also consider their impact on inter-agency coordination and the development of a unified national cybersecurity strategy. While executive orders aim to direct government action, the reality of federal bureaucracy can lead to fragmented implementation. Different agencies may interpret and apply directives in varied ways, leading to overlapping or conflicting regulations. This complexity can be overwhelming for critical infrastructure operators, particularly those that span multiple sectors or jurisdictions. The promise of a cohesive, government-wide approach to OT security, often articulated in presidential directives, can become diluted in the practical realities of agency mandates and resource allocation.

A key myth that presidential directives can perpetuate is the idea that cybersecurity can be effectively managed through top-down mandates alone. True OT security requires a deep understanding of industrial processes, a collaborative relationship between IT and OT teams, and a culture of security awareness that permeates all levels of an organization. Executive orders can mandate policies and procedures, but they cannot easily instill this deeper cultural shift. The emphasis on compliance, driven by the need to demonstrate adherence to executive mandates, can sometimes incentivize organizations to focus on the administrative aspects of security rather than the operational integration of security best practices. This can lead to a situation where security is seen as an add-on, a burden to be managed, rather than an integral component of operational integrity.

The practical impact of these directives often hinges on the quality of their implementation by relevant federal agencies. An executive order directing the Department of Homeland Security (DHS) to enhance information sharing with critical infrastructure operators, for example, will only be effective if DHS has the resources, expertise, and established relationships to effectively engage with a diverse range of private sector entities. If the implementation is characterized by a lack of dedicated OT expertise within the agency, or if the information shared is not tailored to the specific needs of industrial operators, then the directive's intended benefits will not materialize. The myth that an executive order automatically translates into effective action needs to be scrutinized against the realities of bureaucratic implementation and resource constraints.

Ultimately, while presidential directives and executive orders are essential for setting the national agenda on cybersecurity and signaling the importance of protecting critical infrastructure, their impact on genuine OT security is often more nuanced than their lofty pronouncements might suggest. They play a significant role in fueling the "fear machine," driving compliance-based approaches, and sometimes perpetuating the myth that standardized IT security practices can be readily applied to the unique challenges of OT. Achieving true resilience in these vital sectors requires a move beyond these top-down mandates towards a more integrated, risk-based, and culturally embedded approach to security, one that recognizes the distinct nature of OT and prioritizes operational integrity alongside compliance. The challenge lies in ensuring that these powerful policy instruments serve as genuine catalysts for transformative security improvements, rather than merely reinforcing existing narratives and administrative processes.

A Path Forward: Evidence-Based Policy

The persistent reliance on speculative anxieties and commercially driven narratives has demonstrably failed to fortify our nation's operational technology (OT) infrastructure. A paradigm shift is not merely desirable; it is an imperative. This necessitates a foundational recalibration of how we conceive of, develop, and implement cybersecurity policy for these critical systems. The prevailing approach, often characterized by reactive measures and a susceptibility to the sensationalized threat landscape, must yield to a more robust, empirically grounded methodology. This means moving away from policies forged in the crucible of imagined catastrophes and toward frameworks meticulously constructed from the bedrock of observable facts, historical incident data, and a deep, nuanced understanding of the actual operational environments they are intended to protect.

For too long, policy development in OT cybersecurity has operated under a cloud of what can only be described as "epistemic inertia." We have become accustomed to certain tropes, the existential threat posed by nation-state actors, the inevitable "zero-day" exploit that will cripple essential services, the specter of a cascading failure across interconnected industrial control systems (ICS). These are not faceless abstractions; they are the recurring set pieces of named testimony, from Panetta's derailed chemical trains in 2012 to Easterly's "Everything Everywhere, All at Once" in 2024. While they are not entirely without merit, their disproportionate emphasis has created a policy environment that is more responsive to fear than to fact. The challenge lies in disentangling the genuine, albeit often less dramatic, risks from the amplified anxieties that dominate public discourse and, consequently, policy agendas. This requires a deliberate and sustained effort to collect, analyze, and integrate verifiable data into the policymaking process.

Consider the genesis of many OT cybersecurity policies. They often emerge from broad national security concerns, which are then translated into specific directives for critical infrastructure sectors. This top-down approach, while having its place, frequently overlooks the granular realities of how these systems actually function. A policy mandating stringent patching cycles, for instance, might be technically sound from an IT perspective but practically unfeasible or even dangerous in an OT environment where downtime can have immediate physical consequences, or where legacy systems simply cannot accommodate modern patching mechanisms without significant risk of failure. The absence of empirical data on the actual impact of patching failures versus the operational disruption caused by patching attempts leaves policymakers vulnerable to prescribing solutions that are ill-suited to the problem.

The call for evidence-based policymaking is not a call to ignore potential threats. Instead, it is a call to prioritize and address them based on their likelihood, impact, and the feasibility of mitigation strategies. This requires building robust mechanisms for data collection and analysis. We need to systematically gather information on:

Historical Incident Data: A comprehensive, anonymized, and aggregated database of cyber incidents affecting OT systems across all critical infrastructure sectors is essential. This data should go beyond simple breach notifications and examine the root causes, attack vectors, exploited vulnerabilities, the duration of disruptions, and the ultimate impact (e.g., financial losses, physical damage, safety implications). This would allow for the identification of recurring vulnerabilities and the prioritization of mitigation efforts based on actual observed failures. Had such a database existed and been consulted, it is difficult to imagine the catalogue of poisoned cities and derailed chemical trains surviving contact with the empirical record, because that record, even now, contains no events on the mass-casualty, society-crippling scale these scenarios invoke.

Vulnerability Trend Analysis: Beyond known exploits, policy should be informed by a continuous analysis of emerging vulnerabilities specific to OT environments. This includes understanding common misconfigurations, insecure by design features in legacy equipment, and the unique attack surfaces presented by industrial protocols. Such analysis, grounded in real-world penetration testing and security audits of OT systems, would provide a clearer picture of where defenses are most needed.

Effectiveness of Mitigation Strategies: A crucial, yet often overlooked, aspect is the empirical evaluation of the effectiveness of implemented security controls in OT. Policies frequently mandate specific controls (e.g., intrusion detection systems, network segmentation, access control lists). However, there is a significant lack of publicly available data on how well these controls actually perform in diverse OT settings, under real-world attack conditions, and without unduly impacting operational continuity. Evidence-based policy would necessitate pilot programs and ongoing assessments to validate the efficacy and appropriateness of mandated controls.

Operational Constraints and Interdependencies: Policies must be informed by a realistic understanding of the operational constraints inherent in OT environments. This includes system lifecycles that can span decades, the critical need for real-time operations, the high cost and complexity of system upgrades or downtime, and the intricate interdependencies between different industrial processes and physical systems. Without this contextual data, policies risk being theoretical and disconnected from the ground truth.

The current policy landscape, however, is often characterized by a deficit in such empirical grounding. Instead, it leans heavily on frameworks derived from information technology (IT) security, which, as previously discussed, are not always transferable. This reliance is often amplified by commercial interests, where vendors promote solutions tailored to perceived threats, and policymakers, lacking independent data, may be swayed by the loudest or most technologically sophisticated proponents. This creates a self-perpetuating cycle where policy reflects market demands and industry narratives rather than objective risk assessments.

One of the most significant barriers to evidence-based policymaking in OT cybersecurity is the inherent opacity of many industrial operations. Critical infrastructure operators, often private entities, may be reluctant to share detailed incident data due to concerns about liability, reputational damage, or competitive disadvantage. This secrecy, while understandable from a business perspective, creates a significant blind spot for policymakers. Without transparent and aggregated data, the "evidence" available to inform policy remains incomplete, fragmented, and anecdotal. Into that vacuum rushes the very dynamic the University of Utah researchers documented: a discourse that, deprived of disconfirming data, grows steadily more credulous rather than more careful.

To foster a move towards evidence-based policy, several structural changes are necessary. Firstly, there needs to be a greater investment in research and development focused specifically on OT cybersecurity, with a mandate to generate and disseminate empirical findings. This research should be conducted by independent bodies, academic institutions, and government agencies with clear oversight to ensure objectivity. Secondly, regulatory frameworks should be designed to incentivize data sharing, rather than penalize it. This could involve creating safe harbors for companies that report incidents and vulnerability data in good faith, or establishing collaborative platforms where anonymized data can be pooled and analyzed for the collective benefit of all stakeholders.

Furthermore, the focus of policy development needs to shift from a purely compliance-driven model to one that prioritizes resilience and continuous improvement. Compliance, as it is often implemented, focuses on meeting a set of predefined requirements. Evidence-based policy, conversely, would be dynamic and adaptive, informed by ongoing monitoring, analysis, and lessons learned. It would acknowledge that the threat landscape is not static and that security measures must evolve in response to actual observed trends and the effectiveness of implemented controls.

The development of OT-specific risk assessment methodologies, grounded in real-world data rather than hypothetical scenarios, is also paramount. Current frameworks often rely on qualitative assessments that are subjective and prone to bias. An evidence-based approach would integrate quantitative data on incident frequency, impact, and control effectiveness to provide a more objective and actionable understanding of risk. This would allow policymakers to allocate resources more effectively, focusing on the most significant and tractable risks rather than chasing the most sensationalized threats.

Consider the challenge of legacy systems. These systems, often decades old, represent a significant portion of the OT landscape. Policies that mandate the replacement or significant modification of these systems without empirical data on their actual risk profile or the operational consequences of such changes can be counterproductive. Evidence-based policy would involve rigorous analysis of the specific vulnerabilities of these legacy systems, the likelihood of their exploitation, and the practical and economic feasibility of remediation. This might lead to a tiered approach, prioritizing the most critical systems and the most likely attack vectors, rather than a blanket mandate that could cripple operations.

The "fear machine" thrives in the absence of clear, verifiable data. When policymakers are presented with a lack of concrete evidence about the nature and prevalence of OT threats, they are more susceptible to narratives that amplify fear. This can lead to policies that are overly prescriptive, technologically fixated, and ultimately less effective than they could be if they were based on a sober assessment of reality. A commitment to evidence-based policymaking requires a willingness to challenge prevailing assumptions, including the assumptions of the most senior and credentialed voices, and to embrace a more disciplined, data-driven approach. The record set out in this chapter — a "cyber Pearl Harbor" promised in 2012, "something dramatic" declared inevitable in 2014, an "Everything Everywhere, All at Once" scenario invoked in 2024, and none of these catastrophes has materialized at anything approaching the scale invoked — is precisely the kind of evidence that disciplined policymaking would weigh heavily, and that the fear machine has been engineered to ignore.

This transition necessitates a reevaluation of the metrics by which we measure the success of OT cybersecurity policies. Instead of focusing on the number of compliance audits completed or the adoption of specific technologies, we should be measuring improvements in system resilience, reductions in actual incident impact, and the demonstrable ability of operators to recover from disruptions. These metrics, which are inherently evidence-based, would provide a more accurate reflection of the true state of OT security.

In essence, moving towards evidence-based policy in OT cybersecurity is about transforming our approach from one of reactive apprehension to proactive, informed decision-making. It is about building policy frameworks on solid ground, the ground of data, experience, and a realistic understanding of the complex industrial systems we rely upon. This fundamental shift is not merely an academic exercise; it is a critical step in ensuring the actual security and resilience of our nation's most vital infrastructure. The following chapters will examine alternative governance models that can facilitate this transition, moving beyond the limitations of current frameworks and towards a more sustainable and effective future for OT cybersecurity.

Chapter 6

THE MYTH OF IMMINENT CATASTROPHE: A STATISTICAL REALITY CHECK

The prevailing discourse surrounding Operational Technology (OT) cybersecurity often paints a picture of an ever-present, imminent catastrophe, a landscape riddled with sophisticated nation-state actors poised to unleash devastating attacks on critical infrastructure. This narrative, amplified by media sensationalism and vendor marketing, has created a policy environment of alarmism not fully supported by empirical evidence. The previous chapter walked through the incidents themselves, narrating what actually happened at Colonial, at Oldsmar, at the Ukrainian substations, one event at a time. This chapter does something different and, for a book whose thesis is that the industry overstates its evidence, something more dangerous: it counts. A critical examination of OT incident data, the aggregate numbers rather than the individual anecdotes, is essential for understanding the real threat landscape, identifying recurring patterns, and developing policies grounded in statistical reality rather than speculative anxieties. Where Chapter 2 told the stories, this chapter tallies the totals, and the totals tell a story the stories alone cannot.

Analyzing incident data, however, presents a significant challenge. The collection and dissemination of comprehensive, standardized, and anonymized data concerning cyber incidents within OT environments has historically lacked robust infrastructure and consistent reporting mechanisms. Many incidents, particularly those that are minor, quickly remediated, or perceived as not meeting a certain threshold of severity, may not enter the formal reporting channels. Furthermore, proprietary systems and concerns over regulatory compliance, liability, and reputational damage can lead to underreporting by asset owners. This is not a minor caveat, and we will return to it in earnest later in the chapter, because the most serious objection to the argument assembled here is precisely that the numbers undercount. Despite these limitations, several sources provide valuable insights into the nature and frequency of OT cyber incidents. These include government reports from agencies like the Cybersecurity and Infrastructure Security Agency (CISA), reports from private cybersecurity firms that monitor threat intelligence and incident response, and academic research that attempts to aggregate and analyze available data. What follows leans deliberately on the industry's own primary sources, not on outside skeptics, because the most persuasive witness against an overstated threat narrative is the threat industry itself.

What the Fear Machine's Own Numbers Say

Begin with the single most important fact in this chapter, and note where it comes from. Dragos is the leading OT-security vendor in the world; its threat reports are the documents most often cited when a headline warns that industrial control systems are under siege. In its 2024 Year in Review, Dragos states plainly that 2024 industrial ransomware "did not directly impact OT systems." The operational shutdowns that did occur were, in the report's own words, "secondary effects of IT system shutdowns rather than direct OT-targeting exploits." [1] Read that again, because it is the empirical hinge of the entire debate. The fear machine's most-cited primary source concedes that the mechanism is IT-driven. The plant does not go dark because an attacker reached into the controllers and spun the motors to destruction; it goes dark because the company's information technology was encrypted, billing or scheduling or inventory froze, and management pulled the plug as a precaution. That is the same pattern we documented at Colonial Pipeline in Chapter 2, now confirmed as the rule rather than the exception by the industry's flagship dataset.

A 2025–2026 update. The figures above describe the landscape as of 2024. Dragos's more recent year-in-review reports a sharply higher volume of activity — on the order of 119 ransomware groups affecting roughly 3,300 industrial organizations in 2025

— and Dragos's incident responders report observing operational disruption in the operational-technology ransomware engagements they handled that year. Dragos also cautions that many incidents are misclassified as "IT only" because the affected human-machine interfaces, engineering workstations, and historians run Windows. None of this overturns the argument made here: operational disruption still is not catastrophic physical destruction. But the current evidence points to rising operational impact, and the data should be read that way rather than dismissed as mere IT spillover.

Honesty requires the caveat in the same breath as the claim, and the rigor of this book depends on it. The very same Dragos report that concedes ransomware did not touch OT directly also introduces a threat group it designates BAUXITE, credited with genuine programmable logic controller (PLC) compromise capability, and documents new ICS-specific malware families, FrostyGoop and Fuxnet. [2] These are real, they are growing, and they matter. The defensible conclusion is therefore not that OT-targeting attacks are impossible, nor that they will never become common; it is that today they are *rare* and routinely *mislabeled*. The disruptions attributed to "cyberattacks on infrastructure" are, on inspection, overwhelmingly IT compromises whose operational consequences were a downstream side effect, frequently a precautionary choice made by humans rather than a physical effect engineered by malware. Rarity plus mislabeling is a far more modest claim than the impossibility the alarmists accuse skeptics of peddling. It is also the claim the data actually supports.

The point sharpens further when it comes from the very top of the industry. Robert M. Lee, chief executive of Dragos and one of the most credentialed voices in the field, has stated, in a 2023 Dragos briefing: "There have been zero known ICS vulnerabilities ever leveraged in any ICS cyberattack." [3] This is a hostile witness for the alarmist case if ever there was one, the head of the most prominent OT-security vendor, undercutting the very engine that drives much of OT security spending. An entire commercial apparatus is built on the cadence of newly disclosed control system vulnerabilities, the Common Vulnerabilities and Exposures (CVE) advisories that arrive weekly and the patch-or-perish urgency that follows. Yet according to the field's leading practitioner, that catalogue of vulnerabilities has, to date, a record of essentially zero as the leveraged cause of an actual ICS attack. Real attacks on OT, when they happen, tend to use legitimate functionality and stolen or default credentials, an attacker logging in and issuing valid commands, not exotic exploits of unpatched control system flaws. This does not make vulnerability management worthless; it makes the proportion of fear assigned to it badly miscalibrated against the historical record.

The Statistical Reality of OT Incidents

To ground this discussion, consider the typical distribution of publicly reported OT security incidents:

Incident Type	Estimated Frequency	Primary Motivation	Common Impact
Ransomware	High	Financial Gain	Production downtime, data loss, financial cost
Insider Threat (Unintentional)	Moderate-High	N/A (Human Error)	System misconfiguration, temporary outage
Phishing-based Malware	Moderate	Data Theft, Access	Credential compromise, initial network access
Known Vulnerability Exploit	Moderate	Opportunistic Access	System compromise, data exfiltration
Insider Threat (Malicious)	Low-Moderate	Sabotage, Financial	Data theft, process disruption, physical damage
Targeted Nation-State Attack	Low	Espionage, Disruption	Sustained data exfiltration, targeted physical disruption

Note: Frequencies are generalized from public and private sector threat reports. Actual distribution varies by industry.

The shape of that table is the argument in miniature: the rows at the top, financially motivated ransomware and unintentional human error, are tagged "High," while the row everyone fears, the targeted nation-state attack engineered for physical destruction, sits alone at the bottom marked "Low." Examining the attack vectors provides further clarity. Malware, including ransomware and viruses, consistently ranks as a leading cause of OT cyber incidents. Phishing attacks, often targeting individuals within organizations who may have access to OT systems or the networks that connect to them, serve as a primary entry point for malware. This highlights the importance of basic cybersecurity hygiene and user awareness training, even within highly technical OT environments. Exploitation of unpatched vulnerabilities, particularly in legacy systems, is another frequently cited vector. We face the persistent challenge of maintaining OT systems that were not designed with modern cybersecurity in mind and for which patching can be a complex, high-risk operation. The perceived threat of zero-day exploits, while a genuine possibility, appears far less frequently in incident data as a primary cause than the exploitation of known but unaddressed weaknesses or the successful execution of social engineering, a ranking entirely consistent with Lee's observation that no ICS vulnerability is yet known to have been the leveraged cause of an ICS attack.

The types of vulnerabilities exploited are also telling. Often, the vulnerabilities are not in the core industrial control system (ICS) software itself, but rather in the supporting IT infrastructure that has become increasingly interconnected with OT networks. This includes vulnerabilities in operating systems, network devices, and web-facing applications that provide remote access or management capabilities. The blurring of lines between IT and OT networks, while offering operational efficiencies, also expands the attack surface and creates pathways for threats to move from the less secure IT environment into the more critical OT environment. This is precisely the mechanism Dragos identifies, an IT compromise whose effects spill secondarily into operations, and it explains why the honest reading of the data is mislabeling rather than miracle: the attacks are real, but they are IT attacks wearing an OT headline.

The IT-to-OT Attack Path

A common pattern in OT incidents is the initial compromise of IT systems, which then serve as a bridge to the operational network. This is far more frequent than direct attacks on isolated OT components.

Initial Access Vector	Estimated Frequency	Description
IT Network Compromise (e.g., Phishing)	High	Attacker gains access to the corporate network and pivots to the OT network.
Compromised Remote Access	Moderate-High	Exploiting weak credentials or vulnerabilities in VPNs or other remote access tools.
Supply Chain / Third-Party Vendor	Moderate	Compromise through a trusted third-party connection or software update.
Direct OT Network Attack (Internet-facing)	Low-Moderate	Exploiting an OT device (e.g., HMI, PLC) that is improperly exposed to the internet.
Physical Access (e.g., USB drive)	Low	Malicious code introduced directly via physical media.

This reality necessitates robust network segmentation and security controls that acknowledge and mitigate these interconnections. The distribution in the table above is not a rhetorical convenience; it is the operational signature of the Dragos finding. The highest-frequency entry point is an IT network compromise that pivots toward operations, and the lowest-frequency is the direct manipulation of an exposed PLC or human-machine interface, exactly the scenario the catastrophe narrative foregrounds.

Furthermore, analysis of incident data often points to the fact that many compromises are not the result of highly sophisticated, stealthy intrusions. Instead, they are the result of basic security oversights. Weak or default passwords, lack of multi-factor authentication, open network ports that should be closed, and inadequate access controls are frequently found to be contributing factors in successful cyberattacks against OT systems. These are often termed "low-hanging fruit," vulnerabilities that are relatively easy to identify and exploit, and which do not require advanced technical capabilities. The persistence of these basic security failures in reported incidents suggests that a significant portion of the effort to improve OT cybersecurity should focus on addressing these fundamental weaknesses rather than solely on developing defenses against highly advanced threats. The point is not academic: the most-publicized "attacks on water utilities" of recent years involved internet-exposed devices reachable with trivial or default credentials, the digital equivalent of an unlocked door, rather than any novel exploitation of the control logic itself.

The frequency and severity of incidents can also vary significantly by sector, and one statistic in particular has been pressed into alarmist service. IBM's X-Force Threat Intelligence Index has ranked manufacturing the single most-attacked industry, accounting for roughly 26 percent of incidents in 2023 (the year covered by IBM's 2024 index). [4] That number is regularly invoked as evidence that the factory floor, the realm of PLCs and robotic arms and process controllers, is under direct and escalating assault. But the same reporting that produces the ranking also describes the *kind* of attack, and the kind is mundane: credential theft, data extortion, and ransomware, the standard IT menu. Manufacturing tops the chart not because adversaries have learned to manipulate control systems at scale but because manufacturers are exquisitely intolerant of downtime and therefore exquisitely likely to pay. The vulnerability being exploited is legacy IT exposure attached to a business that cannot afford to stop, not the control system itself. Read carefully, "manufacturing is the most-attacked industry" is a statement about extortion economics, not about cyber-physical sabotage, and conflating the two is one of the quieter ways the evidence gets overstated. Healthcare exhibits the same dynamic: connected medical devices and an absolute intolerance for interruption make hospitals prime ransomware targets, but the harm flows from encrypted records and frozen IT, not from an adversary seizing an infusion pump's controls.

It is also crucial to acknowledge the data collection limitations when drawing conclusions. Much of the detailed incident data is proprietary and not publicly disclosed. When data is shared, it is often aggregated, anonymized, or focused on specific types of threats or sectors. This means that the "statistical reality" is often an approximation based on available pieces of a much larger, more complex puzzle. Government agencies are making strides in encouraging reporting and in providing more comprehensive overviews, but a truly complete picture remains elusive. The trends that emerge are nonetheless consistent enough across independent reports to suggest that the narrative of imminent, widespread catastrophe is overstated, and, crucially, the strongest of those sources are the vendors who would profit most from saying otherwise.

The implications of this data-driven perspective for policy are profound. Instead of an approach solely focused on preparing for the worst-case, low-probability nation-state attack, policy should prioritize a multilayered defense strategy that addresses the most common threats effectively. This includes:

1. **Strengthening Defenses Against Common Malware and Ransomware:** Investing in robust endpoint detection and response (EDR) solutions, effective patch management strategies (where feasible and safe), network segmentation, and comprehensive backup and recovery plans are critical. User awareness and training programs to combat phishing and social engineering remain paramount.

2. **Addressing Basic Security Hygiene:** Policies should strongly emphasize and, where appropriate, mandate the implementation of fundamental security controls such as strong password policies, multi-factor authentication, and strict access control management. Regular vulnerability scanning and penetration testing, tailored to OT environments, should be encouraged or required.

3. **Incentivized Incident Reporting:** Creating more robust, secure, and perhaps incentivized mechanisms for asset owners to report cyber incidents and near misses without fear of undue repercussions is an imperative. This would allow for a more accurate understanding of the threat landscape and a more evidence-based approach to policy development. Collaborative platforms for sharing anonymized threat intelligence and best practices can further enhance collective defense, and would directly narrow the undercounting gap discussed below.

4. **Focusing on Resilience and Recovery:** Given that incidents are likely to occur, we must ensure policy strongly emphasizes building resilience into OT systems and developing effective incident response and recovery plans. This includes ensuring business continuity, minimizing downtime, and safeguarding against data loss. Realistic exercises and tabletop simulations are indispensable for testing these plans.

5. **Tailoring Policies to Specific Sectoral Risks:** Recognizing that not all critical infrastructure sectors face the same threats or have the same operational constraints. Policies should be flexible enough to allow for sector-specific risk assessments and tailored mitigation strategies, rather than a one-size-fits-all approach.

6. **Investing in OT-Specific Cybersecurity Research and Development:** Encouraging research into the unique vulnerabilities and defense mechanisms of OT systems, moving beyond IT-centric solutions. This includes developing and validating security tools and strategies that are compatible with the operational demands and legacy nature of industrial control systems.

Counting the Catastrophes: Four in Fifteen Years

If the fear narrative were calibrated to the record, we would expect a long and growing ledger of cyberattacks that produced confirmed cyber-physical effects. We do not have one. We have a list short enough to recite from memory. According to analysis drawing on Dragos and reported by WIRED, in-the-wild malware purpose-built to cause destructive physical sabotage has appeared roughly three times in history: Stuxnet, which damaged Iranian centrifuges; the 2016 Industroyer/CRASHOVERRIDE attack on the Ukrainian grid; and Triton, which targeted a petrochemical safety system in Saudi Arabia. [5] Add the 2015 Ukrainian grid attack, which used legitimate functionality rather than custom destructive malware, and the list reaches about four. Pause on the composition of that list, because it is more deflationary than its length. Two of the four are the *same actor*, Russia, operating in the *same country*, Ukraine, during a period of active geopolitical conflict. The remaining two, Stuxnet and Triton, were the work of nation-states deploying years of effort and resources against single, specific, high-value strategic targets. Roughly fifteen years after Stuxnet announced the cyber-physical era, the independently verified set of cyberattacks with confirmed cyber-physical effects remains small and category-dependent rather than an ever-growing ledger of catastrophe — Stuxnet (physical equipment destruction), the 2015 and 2016 Ukraine grid attacks (service disruption), and TRITON (safety-system targeting, discovered after a fail-safe trip) — with later cases such as FrostyGoop showing a bounded ICS-driven service outage rather than collapse. [6]

Post-2017: capability without catastrophe. The years after TRITON did not produce the cascade of destructive infrastructure attacks the fear narrative predicted, but they did extend the *capability* curve — and the distinction between demonstrated capability and confirmed physical consequence is the one this book keeps pressing. FrostyGoop, malware that speaks the Modbus industrial protocol directly to operational devices, was tied to a heating-service disruption in Lviv, Ukraine, in early 2024 — a real but bounded service outage, not equipment destruction or casualties. PIPEDREAM (also called Incontroller), a modular toolkit built to manipulate industrial controllers, was disclosed by U.S. government agencies and researchers *before* it was deployed against any victim — a capability revealed, not a realized attack. Fuxnet, reported in 2024 against Moscow-area monitoring infrastructure, remains partly self-reported and contested, and its impact should be treated cautiously. Counted honestly, these cases show capability rising while confirmed physical consequence stays rare and bounded — which is this book's argument, not a refutation of it.

Four. That is the number around which an entire industry of imminent catastrophe has been constructed. It is worth stating the obvious caveat in the same breath, as rigor demands: "about four" is a count of *publicly documented, independently verified* incidents, and the qualifier is doing real work, as the discussion of undercounting will make explicit. The list will also lengthen over time; capability is growing, and the PipeDream/Incontroller toolkit disclosed in 2022, together with the FrostyGoop and Fuxnet malware noted earlier, are evidence that the trajectory points upward, not downward. [7] But a trajectory is not a body count, and the body count, after a decade and a half of the most-feared category of attack on earth, is still in the single digits.

The most rigorous attempt to count the right thing reaches the same conclusion from a different direction, and it comes from a respected insider, not a debunker. Dale Peterson, a veteran of the ICS-security community, points to the work of Waterfall Security and ICSSTRIVE, whose annual report is, almost uniquely, an effort to tally incidents with *physical consequences* rather than mere intrusions or data losses. That report logged 57 such incidents worldwide in 2025, and, tellingly, that figure was *down 25 percent* year over year. [8] Fifty-seven is already a small number for an entire planet's industrial base over a full year. But Peterson's decomposition is the part that matters. Strip out the incidents that are ransomware spillover, IT attacks whose effects reached operations secondarily, exactly the mislabeling Dragos describes, and strip out hacktivism, the website defacements and nuisance intrusions that make headlines without touching a process, and the residue of genuine nation-state OT-physical attacks worldwide numbers *three* in 2025, down from six the year before. [9]

Sit with the provenance of that figure. This is not an outside critic straining to minimize the threat. It is a respected practitioner, using the single scariest dataset available, the one report specifically designed to count physical harm, and arriving at single digits, and falling. When the most alarming honest accounting of nation-state cyber-physical attacks on the entire world's infrastructure produces a number you can hold up on one hand, the burden of proof shifts. The question is no longer "why are skeptics downplaying the threat" but "why is the threat being inflated so far beyond what even its most motivated measurers can document."

The Volume Contrast: Thirty Thousand Incidents, No Column for Centrifuges

The scarcity of cyber-physical catastrophe comes into focus only when set beside the staggering volume of cyber activity that is *not* physical at all. Consider Verizon's Data Breach Investigations Report, the most widely cited empirical study of breaches in the field. Its 2024 edition analyzed 30,458 security incidents, of which 10,626 were confirmed data breaches, in a single year. [10] And here is the detail that should reframe the entire debate: the DBIR has no category for physical or process impact. Not because the authors are careless, but because that is simply not where the data lives. The most authoritative breach study on the planet counts more than thirty thousand incidents a year and has no column for "centrifuges destroyed." The data is about credentials, records, money, and access, the IT ledger, not the kinetic one.

The pattern holds across every large-scale measure. The Identity Theft Resource Center has tracked on the order of 60 billion exposed records since it began compiling them in 2005, a number whose scale is almost impossible to picture and whose entire content is data, not damaged machinery. [11] IBM's widely cited 2024 study put the average cost of a data breach at $4.88 million. [12] That figure, updated from the $4.45 million reported the year prior, is the metric the industry reaches for when it wants to convey the cost of an incident, and it is, revealingly, a data-and-dollars metric: lost records, regulatory penalties, notification costs, business interruption measured in revenue. It is never a kinetic figure. There is no line item for a ruptured pipeline or a melted bearing because, in the overwhelming run of cases, there is no rupture and no melt. The definitive breach study counts tens of thousands of incidents a year and has no column for physical destruction, and that single editorial absence tells you more about the true distribution of cyber harm than any vendor threat briefing.

None of this means the dollar figures are trivial, and the contrast cuts both ways. A $4.88 million average breach is a serious event for the organization that suffers it. But it is a business-continuity problem of a recognizable and survivable scale, painful, costly, and bounded, not the trillions-of-dollars, civilization-altering catastrophe theorized for a full-scale cyber-physical war. The honest comparison is between a well-documented annual flood of data breaches averaging a few million dollars apiece and a fifteen-year list of roughly four category-dependent cyber-physical cases, most of them state-on-state acts during armed conflict. The resources are pouring, understandably, toward the data ledger, because that is where the incidents are.

Engaging the Strongest Objection: The Undercounting Argument

A careful reader, and certainly a hostile one, will by now have raised the most serious objection available against everything argued above, and intellectual honesty demands that it be met head-on rather than waved away. The objection is undercounting, and its most persistent advocate is Joe Weiss, a control-systems engineer who has spent years arguing, through his ControlGlobal "Unfettered" blog and elsewhere, that the short lists are short only because we are not looking properly. [13] Weiss's contention runs roughly as follows: the catalogues of "cyber" incidents, the Dragos counts, the Waterfall tally, the four-in-fifteen-years list, capture only the events that announced themselves as cyber, the ones involving recognizable network malware that a security team could detect and attribute. But OT environments, he argues, are precisely the environments least equipped to recognize a cyber cause when one exists. Control systems were engineered for reliability and safety, not forensics; they rarely log the way IT systems log; and when a pump fails or a turbine trips or a process goes out of specification, the default assumption is mechanical fault, sensor drift, or operator error, not a deliberate manipulation. An attacker who alters a process without dropping malware may never be identified as an attacker at all. By this account, the true number of OT incidents with a cyber origin is a dark figure, and the published lists are not ceilings but floors, possibly very low ones.

The first thing to say about this argument is that, on its central methodological point, Weiss is right, and conceding it strengthens rather than weakens the case being made here. OT cyber-forensics genuinely is poor. The instrumentation to distinguish a malicious command from a malfunction frequently does not exist in the field, and the institutional reflex to attribute anomalies to equipment rather than adversaries is real. Every count in this chapter, the Dragos findings, the Waterfall 57, Peterson's three, the four-in-fifteen-years list, should therefore be read as a documented *floor*, an account of what has been verified, not a confident statement of everything that has occurred. This book's entire premise is that the industry overstates its evidence; that premise would be hollow if the book then overstated its own, and so the undercounting point is accepted without reservation. The numbers are imperfect, and they are imperfect in the direction Weiss says.

But conceding the methodology is not the same as conceding the conclusion, and here the argument turns. The relevant question is not whether OT incidents are undercounted in general; it is whether the *specific scenario the fear machine sells*, a remote adversary reaching across a network to manipulate control systems and cause physical harm, is undercounted. And when Weiss's own catalogue of overlooked incidents is examined case by case, most of them turn out not to be that scenario at all. His frequently cited examples include the Maroochy Shire sewage spill in Australia, a serious and genuine act of sabotage, but one committed by a disgruntled former contractor using stolen radio equipment and insider knowledge of the system, a physical insider attack, not a remote network intrusion. [14] Another category he points to involves the diesel emissions "defeat devices," software that manipulated engine controls to cheat pollution tests. That episode was indeed a software manipulation of an industrial control system with real-world physical consequences, but it was *manufacturer fraud*, a company deceiving regulators about its own products, not an external attacker compromising a victim's infrastructure. [15] This is the crucial distinction, and it is the one the fear narrative blurs. Weiss is correct that the universe of "OT incidents with a control-system dimension" is larger than the headline lists, perhaps far larger. But that broader universe is populated overwhelmingly by insiders, by fraud, by accidents, by misconfiguration, and by equipment failures with a software component, the very categories this chapter has emphasized as the real, mundane, high-frequency drivers of OT disruption. It is not populated, except in the same handful of documented cases everyone already cites, by the remote nation-state adversary manipulating controls to wreak physical havoc. Weiss's argument, taken to its logical end, does not resurrect the catastrophe narrative; it reinforces the chapter's thesis by relocating the missing incidents into precisely the categories, insider action and human error, that the alarmist framing ignores. The dark figure is real, but it is dark in the unglamorous direction. To grant the undercounting and still find the remote-sabotage scenario rare is not a contradiction; it is the most defensible position the evidence allows, and it is the one this book takes.

Quantifying Real-World Impact

Quantifying the economic impact of OT cybersecurity incidents is an inherently complex undertaking. Unlike financial services or retail sectors where direct monetary losses from fraud or data theft are more readily calculable, the economic fallout in OT environments often manifests indirectly. Production downtime is a prime example. When a manufacturing plant's industrial control systems are taken offline, even when the cause is an IT compromise rather than direct OT manipulation, the immediate financial implications include lost revenue from unproduced goods. However, the ripple effects can extend far beyond this. There are costs associated with the disruption itself: overtime pay for engineers and technicians attempting to restore operations, expenses for diagnosing and remediating the cyber intrusion, potential penalties for missed delivery deadlines to clients, and the risk of losing those clients permanently. Furthermore, depending on the nature of the manufacturing process, a prolonged shutdown might lead to the spoilage of raw materials or work-in-progress, adding another layer of financial loss.

Consider the case of the Colonial Pipeline ransomware attack of May 2021, examined in detail in Chapter 2 and worth revisiting here only for its economics, because it is the textbook illustration of how an IT compromise produces OT-scale consequences. The malware never touched the operational technology controlling the pipeline; it struck the IT and billing environment, and the company itself made the decision to shut down the largest fuel pipeline on the U.S. East Coast, which supplies roughly 45 percent of the region's fuel. The immediate economic consequence was a severe disruption in fuel supply, leading to panic buying, gasoline shortages, and significant price hikes at the pump across several states. The estimated economic impact was substantial, though precise figures are difficult to isolate from broader consumer behavior, and much of the shortage was driven by panic buying rather than by any actual interruption of pumping capacity. Beyond the direct losses to Colonial itself, including a reported $4.4 million ransom payment (much of which was subsequently recovered by the Department of Justice), the broader damage extended to businesses reliant on fuel deliveries, consumers facing higher transportation costs, and the logistical scramble of rerouting supply. Colonial is, in other words, the perfect specimen of the Dragos finding: a genuine and costly disruption to operations whose root cause and mechanism were entirely on the IT side, the secondary effect of an IT shutdown precisely as the data describes.

Downtime Duration	Frequency in Reported Incidents	Common Causes
Less than 8 hours	High	System misconfiguration, quickly contained malware, minor human error.
8-24 hours	Moderate-High	Ransomware (if backups are available), localized network outage.
1-7 days	Moderate	Widespread ransomware, failure of primary control systems, supply chain disruption.
More than 1 week	Low	Destructive malware (e.g., wiper), significant physical damage, complex recovery.

Note: Durations are estimates based on public incident reporting and can vary significantly by sector and organizational preparedness.

Notice, once more, the shape of the distribution: the most common outcome by a wide margin is a recovery measured in hours, driven by misconfiguration, quickly contained malware, or minor human error, while the rarest outcome, downtime exceeding a week from destructive malware or significant physical damage, sits at the bottom marked "Low." The table is the four-in-fifteen-years finding expressed as a recovery curve. Resilience, not collapse, is the modal result.

The energy sector, particularly electricity generation and distribution, represents another critical area where cyber incidents can have significant economic ramifications. While widespread, prolonged blackouts attributable solely to cyberattacks have been rare, and the documented record bears this out, localized disruptions and the costs associated with preventing them are very real. A cyber intrusion targeting the operational systems of a power substation could necessitate its temporary isolation, leading to power outages in a specific geographic area. The economic impact here would include the direct costs of restoration, potential damage to sensitive electrical equipment if not properly managed, and the broader economic losses for businesses and residents in the affected area. The cost of cybersecurity investments made by utilities to prevent such incidents, while a necessary expenditure for risk mitigation, also represents a significant economic factor, often borne by consumers through their electricity bills. These are not hypothetical costs; they are real, ongoing investments driven by the perceived risk, even where large-scale catastrophic events remain, by the numbers just rehearsed, infrequent. The two documented attacks that did interrupt power, both in Ukraine (2015 and 2016), restored customer power within hours—largely through manual operator intervention—though full system recovery took considerably longer, a point developed below.

Beyond direct financial losses and lost revenue, the impact on OT systems can also manifest as physical damage. This is the most concerning category of impact because it directly affects the integrity of industrial processes and machinery, often leading to costly repairs, prolonged downtime, and in the most severe cases, safety hazards. It is also, as the count above established, the rarest. The Stuxnet worm, discovered in 2010, is the foundational example and the first entry on the list of four. It was a sophisticated state operation widely reported to have been a joint U.S.–Israeli effort built to sabotage Iran's nuclear program by targeting specific centrifuges; it exploited the Siemens Step7 software used to control the programmable logic controllers driving those centrifuges, subtly altering their speed so they would damage themselves while reporting normal operation to their operators. Stuxnet proved that cyber action *can* cause physical destruction. What the subsequent fifteen years proved is how seldom that capability is realized in the wild, and at what cost it must be assembled: years of work, multiple zero-day exploits, intelligence on an air-gapped facility, and the resources of a nation-state, all spent against a single high-value target.

The economic cost of physical damage, when it does occur, can be immense, often far exceeding the cost of a simple software remediation. It involves not only the replacement of damaged hardware but also extensive safety investigations, potential environmental cleanup where hazardous materials are involved, and the long lead times associated with specialized industrial equipment. The reputational damage and potential legal liabilities stemming from an incident that harms personnel can be catastrophic for an organization. This is precisely why the category warrants serious attention even at low frequency: the expected cost of a risk is its probability multiplied by its magnitude, and physical-damage events carry enormous magnitude. But that same logic is what the fear machine exploits, foregrounding the magnitude while quietly omitting the probability. An honest risk assessment has to hold both numbers at once, and the probability term, on the evidence, is small.

Service disruptions in OT environments are a frequent occurrence, though their scale and severity are generally more contained than the apocalyptic visions presented by some alarmists. In the healthcare sector, a cyberattack on a hospital's IT network can have immediate implications for connected medical devices: a ransomware attack that encrypts patient records can disrupt the functionality of devices that rely on those records, such as infusion pumps, diagnostic imaging equipment, or patient monitoring systems. The impact is a loss of critical functionality, potentially delaying treatment, and the consequences for patient care can be severe. But note the mechanism: the harm again flows from the IT compromise and the loss of supporting data, not from an adversary directly commandeering the device's controls. The economic impact includes incident response, the need to divert patients to other facilities, and lost revenue during periods of reduced capacity, a profile that maps onto the IBM breach-cost data far more naturally than onto any model of kinetic sabotage.

In the water and wastewater sector, cyber incidents can in theory lead to disruptions in the operation of pumps, treatment facilities, and distribution networks, the manipulation of chemical dosages, the disruption of filtration, or the shutdown of pumping stations. Reports and advisories from agencies such as CISA have highlighted instances where attackers gained access to water utility control systems. Yet widespread contamination or total loss of service from a cyberattack has not materialized, and the documented intrusions, on inspection, have overwhelmingly been the low-hanging-fruit variety: internet-exposed control devices reachable through weak or default credentials, often resulting in nuisance defacement rather than any manipulation of water chemistry. The harm in such cases is real but bounded, and the safeguards that limit it, alarms, lag times between a setpoint change and a physical effect, and human operators positioned to notice and intervene, are exactly the defenses the catastrophe narrative tends to leave out of its scenarios.

Comparing these documented impacts to the hypothetical projections is crucial for a balanced risk assessment. Vendors frequently posit scenarios in which a single sophisticated nation-state actor induces widespread systemic failure by crippling critical infrastructure. While the *potential* for serious disruption is genuine, the *frequency* and *mechanisms* of these maximalist claims diverge sharply from observed reality. The empirical data, as assembled in this chapter, suggests that actual impacts are more localized, more often rooted in IT compromise or human error than in direct OT manipulation, and far more frequently characterized by recoverable disruption and financial loss than by societal collapse. Ransomware, vastly more common than targeted state sabotage, aims for financial gain, and its impact, while devastating for the affected entity, is usually contained to that entity and its immediate supply chain. The catastrophe narrative also routinely overlooks the resilience built into critical infrastructure, the redundancy of operational processes, and the role of human operators as a final line of defense able to intervene when automated systems behave erratically.

The economic losses associated with these common incidents, while substantial, generally fall within the realm of manageable, if painful, business-continuity challenges rather than existential threats to national security. The IBM 2024 figure of $4.88 million (IBM's 2025 report put the global average at about $4.44 million) as the average cost of a data breach provides a useful benchmark for the magnitude of financial consequences. While that number primarily reflects IT-related breaches, and OT costs can run higher where physical damage or extended downtime is involved, it remains a data-and-dollars figure, distinct by orders of magnitude from the trillions of dollars in losses routinely theorized for a full-scale cyber war. The focus on quantifying these impacts is not to downplay the seriousness of OT cybersecurity threats, but to ensure our understanding and response are proportionate to the actual risks. Policies, investments, and public discourse should be informed by a clear-eyed assessment of what has happened, what is happening, and what is most likely to happen, rather than driven solely by the most dramatic, least probable worst-case scenarios. This is a shift from a reactive posture fueled by fear of the unknown to a proactive strategy grounded in the observable realities of the threat landscape, a transition from hypothetical catastrophe to statistical reality that is not about complacency but about clarity, focus, and efficiency in safeguarding the systems that underpin modern society.

Human Error: The Higher-Frequency Threat

The pervasive narrative of imminent catastrophic cyberattacks on operational technology, while dramatic, often overshadows a more mundane yet equally impactful culprit: human error. While the specter of sophisticated nation-state actors or highly organized criminal groups launching devastating cyber offensives is a compelling concern, empirical evidence increasingly points to operational mistakes, inadequate training, and procedural lapses as more frequent and significant drivers of OT system failures and disruptions. This is also where, as the undercounting discussion established, the genuinely overlooked incidents predominantly reside, in the unglamorous categories of insider action and operator mistake rather than in remote sabotage. This subsection examines the statistical reality of human error in OT environments, presenting data and patterns that underscore its disproportionate influence on system vulnerabilities and the disruption of critical services.

Consider the sheer complexity and the demanding operational tempo inherent in many OT environments. These systems, responsible for managing physical processes in real time, often operate under stringent uptime requirements and in environments that can be physically challenging or hazardous. Human operators are tasked with monitoring, controlling, and maintaining these intricate systems, often in shifts that can extend for many hours. The potential for error, fatigue, distraction, or misinterpretation is, therefore, an intrinsic aspect of human involvement. Unlike IT systems, where a mistake might lead to a data breach or a system slowdown, an error in an OT environment can have immediate physical consequences, ranging from equipment malfunction to significant safety incidents.

Data collected from various industrial sectors consistently highlights the role of human factors in operational incidents. Analyses of industrial accidents often reveal that while technological failures may be a contributing factor, the root cause frequently traces back to human actions or inactions. This can manifest in several ways: incorrect configuration of control parameters, failure to follow established shutdown or startup procedures, misinterpreting sensor readings, or inadvertently executing the wrong command. In some cases, these errors are borne out of insufficient training. Operators may not fully grasp the intricacies of the systems they manage, the potential consequences of certain actions, or the proper protocols for responding to abnormal situations. The rapid pace of technological change in OT, coupled with the inherent difficulty in finding and retaining highly skilled personnel, exacerbates this training deficit. Moreover, adherence to procedures is paramount in OT environments. Standard operating procedures (SOPs) are designed to ensure safe, efficient, and reliable operation. In practice, however, there can be a tendency for operators to deviate from these procedures, driven by a perceived need to expedite a process, a misunderstanding of the procedure's importance, or simply habit. While a minor deviation might go unnoticed, a more significant breach of protocol can have cascading effects. Skipping a crucial safety check before a maintenance operation on a high-pressure pipeline, for instance, could lead to catastrophic failure, and the economic impact of such an event, encompassing repair costs, lost production, environmental remediation, and potential fines, can far outweigh the cost of diligent procedural adherence.

The patterns are concrete and sector-specific. In manufacturing, a common point of failure involves the programming or configuration of programmable logic controllers, the workhorses of industrial automation that control everything from conveyor belts to robotic arms. An error in PLC code, introduced during a routine update or modification, can produce unexpected machinery behavior ranging from a machine stopping prematurely to a robotic arm moving erratically and damaging itself or surrounding equipment. Investigations routinely trace such errors to incomplete testing of code changes, a lack of proper version control, or inadequate understanding by the implementing engineer. The cost of troubleshooting and rectifying these errors, to say nothing of downtime and equipment damage, can be substantial, and none of it requires an adversary at all.

In energy production and distribution, human error can carry particularly significant consequences. Switching between power sources or rerouting electricity flow within a grid are complex operations requiring precise timing and careful coordination; a misinterpreted signal or a step executed out of sequence can lead to voltage instability, equipment overload, or localized blackouts. While cyberattacks are a known threat, historical analyses of power outages consistently reveal that operational errors, faulty maintenance practices, and incorrect switching operations have been responsible for a considerable share of disruptions, a far larger share, by the numbers, than the two cyberattacks ever documented to have interrupted power, both in Ukraine. The economic fallout from even a localized outage can be significant, affecting businesses, residents, and critical services within the affected area.

The water and wastewater treatment sector, vital for public health, is likewise susceptible to human-induced failures. The operation of pumps, chemical dosing systems, and filtration units requires constant monitoring and precise adjustment. An operator error, such as miscalculating the amount of disinfectant added to a water supply or failing to monitor tank levels correctly, can lead to compromised water quality or environmental contamination. The outcomes a malicious actor might aim for can thus be produced, far more commonly, by simple errors of judgment or procedure, and here the undercounting argument is most instructive: the Maroochy Shire spill, frequently filed under "OT cyber incident," was the work of a disgruntled former contractor using stolen radio equipment, a category much closer to the human-factors failures described in this section than to the remote-attacker scenario the fear narrative sells. The cost of remediation, regulatory fines, and the erosion of public trust can be immense.

The healthcare sector presents a domain where human error in OT systems can have life-or-death consequences. Modern hospitals rely on a vast array of interconnected medical devices, including infusion pumps, ventilators, patient monitoring systems, and diagnostic imaging equipment. While cybersecurity vulnerabilities are a growing concern, the potential for human error in the operation and maintenance of these devices is also a significant risk. An incorrect setting entered into an infusion pump, a mistake in connecting tubing, or a failure to calibrate equipment properly can lead to patient harm. These are not "cyber" errors in the sense of external malicious code, but they are errors in the operation of networked, automated systems that demand rigorous protocols, extensive training, and a culture of safety to mitigate. Investigations into adverse patient events frequently identify human error as a primary contributing cause, even where underlying system vulnerabilities exist.

The emphasis on human error is not intended to diminish the threat posed by sophisticated cyberattacks. Rather, it seeks to provide a more balanced perspective on the sources of risk in OT environments. Many organizations, particularly small and medium-sized enterprises, lack the resources or expertise to defend against advanced persistent threats, yet they are far more likely to experience disruptions stemming from internal operational issues. Investing in comprehensive training programs, fostering a strong safety culture, implementing robust procedural controls, and ensuring clear lines of accountability are often more cost-effective and impactful strategies for reducing operational disruptions than focusing solely on cutting-edge cyber defenses against an adversary who, statistically, is far less likely to arrive than a tired operator on a double shift.

Furthermore, the concept of "human error" itself is nuanced; it is not always about individual negligence. Often, systemic issues contribute to errors: poorly designed interfaces prone to misinterpretation, inadequate staffing levels leading to operator fatigue, a lack of clear communication channels, or a work environment that discourages reporting of near misses. Addressing these underlying systemic factors is crucial for creating a truly resilient OT environment. Organizations must move beyond blaming individuals and focus on creating systems and processes that minimize the likelihood of error and provide safety nets when errors do occur, through sound human-machine interface design, adequate rest periods, robust quality assurance for system changes, and a culture where safety concerns can be raised without fear of reprisal.

The proliferation of legacy systems compounds the problem. Many critical infrastructure systems rely on equipment that is decades old, designed long before modern cybersecurity threats were conceived. While these systems may be rugged and reliable, their interfaces and operational paradigms can be less intuitive and more prone to operator misinterpretation, and their documentation may be incomplete or outdated, making it harder for new operators to learn them fully. The temptation to cut corners or rely on informal workarounds rises in such environments, increasing the likelihood of error. The rapid pace of modernization presents its own challenges: as new technologies are introduced, operators must be trained to use them effectively and safely, and a hasty or inadequate transition, the rollout of a new supervisory control and data acquisition (SCADA) system, for instance, can introduce complexities that operators are not prepared to handle, producing a surge in operational errors. The interconnectedness of modern OT systems means that an error in one component can ripple across the entire operation: a minor misconfiguration in a sensor network can feed incorrect data into a control system, triggering a cascade of incorrect actions by other automated equipment. This interconnectedness, while enabling efficiency, amplifies the potential impact of any single point of failure, including those stemming from human error.

In sum, while sophisticated cyberattacks remain a concern, the evidence strongly suggests that human error is a more pervasive and immediate threat to the reliable operation of OT systems. The financial and operational consequences of misconfigurations, procedural violations, inadequate training, and systemic weaknesses are significant and ongoing. Acknowledging and proactively addressing these human factors is not a distraction from cybersecurity; it is a fundamental component of building robust and resilient critical infrastructure. By prioritizing operator training, procedural discipline, intelligent system design, and a culture that values safety and learning, organizations can significantly reduce the risk of disruptions, a strategic prioritization of resources toward the most common and impactful causes of system failures in the real world.

The Limited Scope of Documented Attacks

One crucial observation from the figures assembled in this chapter is that even when sophisticated attacks occur, they rarely achieve the catastrophic, widespread impact that vendor projections often suggest.

When such events *do* occur, their consequences are frequently contained. This containment is not accidental; it stems from critical factors that current risk assessments routinely overlook while fixating on hypothetical maximum damage. These de facto safeguards include operational resilience, inherent system redundancy, and the formidable difficulty of orchestrating the precise confluence of factors needed for a truly catastrophic outcome. An attack aimed at disrupting a power grid might target the SCADA systems controlling a specific substation; while this can cause localized outages and significant inconvenience, it rarely produces a nationwide blackout. Grids are designed with decentralization and manual override capabilities, so other sections of the network can compensate and human operators can intervene to restore service. The "domino effect" that leads to widespread collapse is harder to engineer than theoretical models suggest, requiring deep, precise knowledge of specific network configurations and a sustained, coordinated effort across multiple nodes.

The capabilities and motivations of adversaries are also more specific and less destructive than the narrative of state-sponsored digital warfare suggests. While nation-states are certainly capable of developing advanced cyber tools, their primary objectives in targeting OT are seldom aligned with causing maximum physical destruction for its own sake. Espionage, intellectual property theft, and the disruption of specific industrial processes for economic advantage are more common goals. Even where physical disruption is a stated aim, execution can be hampered by a lack of detailed, current knowledge of the target's specific architecture, operational procedures, and human factors. This is consistent with the economics noted earlier: the ransom-seeking criminal has every incentive to keep an asset intact and billable, not to destroy it. The digital landscape of OT is incredibly diverse, with bespoke systems, legacy equipment, and unique configurations making a "one-size-fits-all" attack strategy highly ineffective. The reconnaissance and engineering required to tailor a sophisticated attack to a specific OT environment is immense, a significant barrier for all but the most determined and well-resourced adversaries, which is precisely why the verified list of such successes remains at roughly four.

Consider Stuxnet once more, not to re-narrate it but to read it as a data point about scale. Highly advanced as it was, its impact, though significant for the specific centrifuges it targeted, was ultimately contained to a particular uranium enrichment facility. It did not cascade to cripple Iran's entire nuclear program or destabilize regional power grids. The attack was narrowly targeted, requiring delivery via removable media into an air-gapped environment and a deep understanding of specific Siemens PLC controllers and the physical processes they managed. That specificity, while making the attack devastatingly effective against its intended target, also limited its broader applicability and, crucially, its potential for widespread damage beyond that facility. The effort involved in developing and deploying Stuxnet was colossal, which is exactly why such attacks are reserved for specific, high-value strategic objectives rather than indiscriminate destruction, and exactly why they appear roughly once every few years rather than weekly.

The Ukrainian grid attacks of 2015 and 2016, attributed to the Russian GRU-linked Sandworm group and accounting for half of the four-incident list, tell the same story of containment. They did cause significant outages affecting hundreds of thousands of people, but even these were not civilization-ending events. The outages were temporary, lasting hours in the affected regions, and the systems were brought back online, frequently through manual intervention and preexisting contingency plans. The attackers exploited specific weaknesses in SCADA systems, but the resilience of the broader grid and the rapid response of human operators were decisive in limiting the duration and ultimate impact. These attacks demonstrate a real capability for targeted disruption, and they should be granted in full, but they also demonstrate, with equal clarity, the role of human operators and robust recovery procedures in mitigating the worst outcomes. They represent a step toward widespread disruption, not the realization of it, and they remain, years later, the high-water mark rather than the routine case.

The resources required to mount and sustain truly catastrophic cyberattacks on OT are immense, often exceeding the capabilities of even well-funded non-state actors. The development of custom malware, the identification of zero-day vulnerabilities, the penetration of air-gapped or highly segmented networks, and the coordination of multiple attack vectors all demand a level of technical expertise, financial backing, and strategic planning that is rare. When these resources are deployed, they are often directed toward objectives that do not involve rendering entire nations or critical sectors perpetually inoperable. The economic incentives for cybercriminals, again, lean toward ransomware and data theft rather than the destruction of physical assets, which would devalue those assets and reduce the potential for financial gain. This is the structural reason the IBM and Verizon datasets bulge with data breaches while the cyber-physical-effects ledger remains small and category-dependent.

The persistent focus on theoretical "mega-attacks" warps the budget toward the spectacular and away from the probable. Organizations may invest heavily in defenses against highly sophisticated, low-probability threats while underfunding basic cybersecurity hygiene, employee training, and robust incident response planning, the measures that address the far more common and impactful threats such as human error or opportunistic malware. The narrative of imminent catastrophe can create a sense of urgency that drives immediate, often expensive, security purchases, sometimes without a clear understanding of the actual risks faced by a particular organization. This is not to say sophisticated attacks are impossible or should be ignored; the documented growth in OT-specific capability forbids such complacency. It is to say that their probability and potential impact in the real world are routinely exaggerated, producing a skewed perception of risk that the numbers in this chapter are meant to correct.

Furthermore, the concept of OT "failure" is itself complex. An industrial process might be halted by a cyberattack, but it might equally be halted by equipment failure, by power outages unrelated to cyber activity, by supply chain disruptions, or by human error. Differentiating the cause of an operational disruption can be genuinely difficult, and a cyberattack might exploit preexisting weaknesses that were not primarily cyber-related, the very ambiguity Joe Weiss rightly emphasizes. That ambiguity cuts in both directions: it means some cyber causes go unrecognized, and it means some non-cyber failures get attributed, in the retelling, to cyberattack. An honest accounting acknowledges both error bars rather than assuming the uncertainty runs only in the alarming direction.

The reality of cyber conflict in the OT space is more often characterized by a spectrum of activities, from low-level probing and reconnaissance to opportunistic malware infections and targeted but contained disruptions. The "catastrophic" attacks that dominate headlines are, by historical evidence, exceedingly rare, roughly four in fifteen years by the destructive-malware count, and three worldwide in the most recent year by Peterson's nation-state decomposition. This does not diminish the importance of cybersecurity in OT; it suggests that focus and resources should be calibrated to the actual observed threat landscape, which is dominated by less sophisticated threats and operational vulnerabilities. Adversary capabilities are evolving, and the potential for more devastating attacks certainly exists, the trajectory of PipeDream, FrostyGoop, Fuxnet, and BAUXITE (as catalogued by Dragos and others) makes that undeniable. But the historical record provides a valuable counterpoint to the most alarmist predictions, suggesting that resilience, human intervention, and the sheer difficulty of achieving widespread catastrophic impact often serve as bulwarks against the most extreme scenarios. Understanding this statistical reality is crucial for developing effective, proportionate, and sustainable OT security strategies.

A pragmatic, data-informed approach is more sustainable and effective in safeguarding the critical services OT underpins. It shifts focus from preparing for a hypothetical apocalypse to building robust defenses against the probable, and often equally damaging, everyday threats.

The prevalent discourse surrounding cyber threats to operational technology frequently amplifies the specter of catastrophic, civilization-altering events. These narratives, while effective in galvanizing attention and securing resources, paint a picture that deviates from the documented empirical evidence of OT cyber incidents. A rigorous examination of historical data and actual events reveals a more granular landscape of threats, where the probability and impact of malicious cyber activities are more contained and less apocalyptic than commonly portrayed. This divergence between narrative and reality, a divergence the industry's own primary sources confirm, necessitates a critical reevaluation of how we prioritize cybersecurity within OT environments. The current tendency to focus disproportionately on highly sophisticated, low-probability, high-impact scenarios, often termed "mega-attacks," risks diverting valuable resources away from more pervasive and demonstrably impactful threats.

When we move beyond the sensationalized hypothetical scenarios and examine the actual documented cyber incidents affecting OT systems, a compelling argument emerges for recalibrating threat prioritization. The focus should shift from the rare, albeit devastating, nation-state-sponsored attacks or sophisticated criminal endeavors that aim for widespread physical destruction, toward a more pragmatic, data-driven approach. This recalibration suggests redirecting significant policy attention and investment away from speculative, low-probability catastrophic events and toward the more frequent, tangible, and often mundane risks that pose a consistent and significant threat to industrial operations: human error, insider threats, and the exploitation of fundamental system vulnerabilities arising from systemic weaknesses in governance, management, and basic cybersecurity hygiene.

Consider once more the practicalities and resource demands of executing a truly catastrophic cyberattack on large-scale OT infrastructure such as a national power grid or a critical water treatment facility. Such an endeavor would require an extraordinary level of sustained effort: deep reconnaissance of the target's unique architecture, the development of bespoke malware capable of evading multiple layers of detection, the identification and exploitation of zero-day vulnerabilities, and the intricate coordination of numerous attack vectors. Its success would further depend on a confluence of factors, precise timing, the absence of effective human intervention, and the exploitation of cascading vulnerabilities that are extraordinarily difficult to predict and engineer at scale. The historical record bears this out: the instances where such comprehensive and devastating outcomes have been achieved are exceedingly rare and confined to highly specific, targeted environments rather than broad-spectrum destruction, four in fifteen years, two of them the same actor in the same country.

The resources, both human and financial, required for these "mega-attacks" are substantial, often placing them beyond the reach of all but the most well-resourced state actors. Even for such actors, the strategic calculus of deploying potent and potentially detectable tools might favor more discreet methods of achieving geopolitical or economic objectives. Cyber espionage, intellectual property theft, or the subtle manipulation of industrial processes for economic leverage might offer a higher return on investment with a lower risk of attribution and retaliation. The notion of indiscriminate, widespread destruction for its own sake, while chilling, appears to be a far less frequent motivation for sophisticated adversaries than targeted, calculated objectives, which is exactly what Peterson's count of three nation-state OT-physical attacks worldwide in a year would lead one to expect.

Conversely, the spectrum of cyber threats that demonstrably impact OT systems on a regular basis is often far less technologically advanced, yet remarkably effective. Human error stands out as a perennial and significant contributor to operational disruptions. Accidental misconfigurations, inadvertent execution of malicious files, the use of unsecured portable media, or simple failures to follow established security protocols can inadvertently open doors for attackers or directly cause system failures. The human element, often hailed as the first line of defense, can also be the weakest link, and its impact is far more widespread and predictable than that of a highly sophisticated nation-state attack. Training deficiencies, fatigue, and a lack of security awareness create fertile ground for exploitation, frequently by unsophisticated means.

Insider threats, whether malicious or unintentional, represent another class of risk that warrants greater attention than it typically receives. An insider, by definition, has legitimate access to systems and facilities, significantly lowering the barriers to causing disruption or compromise. Disgruntled employees, individuals seeking financial gain, or those acting under duress can leverage existing privileges to wreak havoc. Consequences range from data exfiltration and intellectual property theft to the manipulation or destruction of critical operational processes. Unlike external attackers who must overcome perimeter defenses and navigate complex network architectures, insiders operate within the trusted zone, making their actions harder to detect and mitigate through traditional security measures. As the undercounting discussion made clear, when overlooked OT incidents are finally attributed, they land disproportionately in this category, Maroochy Shire being the canonical example, rather than in the remote-sabotage column. The probability of such an event within any given organization is arguably higher than that of a sophisticated state-sponsored attack.

Moreover, the persistent neglect of fundamental cybersecurity hygiene and robust governance frameworks in many OT environments creates a landscape ripe for exploitation by less sophisticated threats. Outdated software, unpatched vulnerabilities, weak authentication mechanisms, lack of network segmentation, and inadequate physical security measures are common deficiencies. These are not the vulnerabilities that advanced persistent threats exclusively exploit; they are the low-hanging fruit that opportunistic malware, less sophisticated criminal groups, and even automated scanning tools can readily identify and leverage. The SolarWinds incident, while involving sophisticated attack vectors, also highlighted how the compromise of a widely used software supply chain could have far-reaching consequences, affecting numerous organizations that might otherwise have had robust defenses against direct attacks. This underscores the interconnectedness of IT and OT security and the importance of addressing foundational vulnerabilities across the entire digital ecosystem, the same IT-to-OT pathway the Dragos data identifies as the dominant mechanism of OT disruption.

That same tendency drains more than budgets; it drains attention and scarce talent. Organizations may find themselves investing heavily in advanced threat detection systems designed to catch the most elusive nation-state actors while underfunding essential programs like comprehensive employee training, rigorous patch management, regular vulnerability assessments, and well-rehearsed incident response plans. These latter measures, while seemingly less glamorous, address the more frequent and impactful threats that organizations actually face day to day. A well-trained workforce, diligent patching, and a clear, practiced incident response capability can significantly reduce the likelihood and impact of a wide range of cyber incidents, from ransomware attacks to accidental system misconfigurations.

The myth of imminent catastrophe can also foster a sense of panic that drives hasty and often inefficient security decisions. Faced with sensationalized reports and urgent warnings, organizations might procure expensive security solutions without a thorough risk assessment, leading to investments that do not align with their actual threat profile. This reactive approach is a far cry from a strategic, risk-based security posture. A data-driven approach to threat prioritization, informed by actual incident data, vulnerability assessments, and a clear understanding of the organization's specific operational context, is far more likely to yield effective and sustainable security outcomes. This means understanding which vulnerabilities are most likely to be exploited within a given environment, what types of adversaries are most likely to target the organization, and what the most probable consequences of a successful attack actually are, questions whose answers, on the evidence of this chapter, point toward IT compromise and human error far more often than toward kinetic sabotage.

The difficulty in achieving truly catastrophic outcomes in OT environments should not be misconstrued as an argument for complacency. The potential for severe damage is undeniably real, and the consequences of a successful attack on critical infrastructure could be profound. However, the statistical reality suggests that such extreme events are not the norm but the exception, often requiring a perfect storm of technical sophistication, precise targeting, opportune timing, and a degree of systemic fragility that may not be present in all or even most OT deployments. The resilience inherent in many industrial control systems, the presence of manual overrides, the redundancy of certain components, and the proactive intervention of skilled human operators often serve as crucial buffers against the worst-case scenarios.

Therefore, a pragmatic approach to OT cybersecurity demands a shift in focus. The policy landscape and industry best practices should evolve to reflect the observable patterns of cyber threats rather than being driven solely by the most extreme, yet least probable, scenarios. This means investing in the fundamentals: robust access controls, comprehensive asset management, regular security awareness training, diligent vulnerability management, and well-defined and regularly tested incident response and disaster recovery plans. It means acknowledging the significant role of human factors and insider threats and developing strategies to mitigate them. It also means fostering a culture of security that permeates all levels of an organization, from the shop floor to the executive suite.

By recalibrating threat prioritization, organizations can allocate their limited resources more effectively, focusing on the threats that pose the most consistent and demonstrable risk. This does not mean abandoning defenses against sophisticated attacks altogether, but ensuring that the foundational elements of cybersecurity are firmly in place and that the response to threats is proportionate to their actual likelihood and potential impact. This data-informed, risk-centric approach is more sustainable, more efficient, and ultimately more effective in safeguarding operational technology and the critical services it underpins. The narrative of imminent catastrophe, while serving a purpose in raising initial awareness, should not dictate long-term security strategies. Instead, the empirical evidence of what actually happens in the realm of OT cyber incidents, thirty thousand data breaches a year and no column for centrifuges, roughly four category-dependent cyber-physical cases in fifteen years, three nation-state OT-physical incidents worldwide in the most recent year and falling, and the industry's own leading vendor conceding that ransomware did not directly touch OT, should guide our efforts toward building truly resilient and secure systems. The focus should lie not on preparing for the unlikely apocalypse, but on building robust defenses against the more probable, and often equally damaging, everyday threats.

Chapter 7

INTERNATIONAL PERSPECTIVES: DIVERSE APPROACHES TO OT SECURITY

Israel: Centralization, Pragmatism, and Resilience

Israel's approach to cybersecurity, particularly concerning its Operational Technology (OT) landscape, offers a compelling case study in centralized coordination and pragmatic risk management. The establishment and evolution of the Israel National Cyber Directorate (INCD) are central to understanding this strategy. Unlike some nations that have adopted a more fragmented or decentralized approach, Israel has opted for a robust, top-down model, consolidating cyber defense responsibilities under a single, powerful entity. This centralization is a strategic imperative for a coherent national response to complex cyber threats. The INCD, established in 2018 through the merger of the National Cyber Bureau and the National Cyber Security Authority, serves as the principal government agency responsible for developing and implementing Israel's national cyber policy, a mandate that explicitly includes the protection of critical infrastructure and, by extension, its embedded OT systems.

The INCD's mandate is broad, encompassing the development of national cyber defense capabilities, the promotion of cyber awareness and education, and the coordination of incident response across all sectors of the economy. This unified structure allows for a more efficient allocation of resources, a streamlined decision-making process, and a consistent application of security policies. In the context of OT security, this means that the INCD can foster a unified understanding of risks, promote standardized security practices, and ensure that the lessons learned from one sector or incident are rapidly disseminated and applied across others. This is particularly crucial for OT environments, which often operate with unique technical constraints, legacy systems, and a distinct operational logic that differs significantly from traditional Information Technology (IT) environments. The INCD's role is to bridge this gap, ensuring that OT security is not an afterthought but an integral component of national cybersecurity strategy.

A hallmark of the INCD's strategy is its decidedly pragmatic focus. While parts of the global cybersecurity discourse can gravitate towards speculative, high-impact, low-probability scenarios — the 'mega-attacks' that capture headlines — Israel's approach appears more grounded in the observable realities of cyber threats affecting industrial control systems and other OT assets. This pragmatism manifests in several ways. Firstly, there is a strong emphasis on building resilience rather than solely on prevention. While preventing breaches is a key objective, the INCD's posture reflects the recognition that in complex and interconnected OT environments, complete prevention can be an elusive goal. Therefore, significant effort is directed toward ensuring that systems can withstand, respond to, and recover from cyber incidents with minimal disruption. This involves developing robust incident response plans, implementing effective backup and recovery mechanisms, and fostering a culture of continuous monitoring and adaptation.

Secondly, the INCD's pragmatic approach prioritizes tangible risks that have a demonstrable impact on operations. This includes addressing vulnerabilities that are frequently exploited, such as those arising from human error, inadequate patching of systems, the use of insecure legacy equipment, and the inherent complexities of integrating IT and OT networks. The focus is less on the theoretical capabilities of sophisticated nation-state adversaries (though these are not ignored) and more on the practical security measures that can effectively mitigate the everyday threats that industrial operators face. This could involve advocating for stronger access control mechanisms, implementing network segmentation to limit the lateral movement of threats, enhancing the security of remote access solutions, and ensuring that personnel are adequately trained to recognize and respond to cyber threats. The INCD's guidance and regulatory framework tend to reflect these on-the-ground realities, providing actionable advice and requirements that are directly applicable to OT environments.

This pragmatic orientation also extends to the INCD's engagement with the private sector. Recognizing that a significant portion of critical infrastructure is privately owned and operated, the INCD has actively sought to foster collaboration and information sharing with industry stakeholders. This is achieved through various mechanisms, including working groups, information sharing platforms, and joint exercises. The aim is to create a symbiotic relationship where the government provides strategic direction, regulatory oversight, and threat intelligence, while the private sector provides operational expertise, industry-specific knowledge, and the implementation of security measures. This collaborative model is crucial for OT security, as the operational realities and risk landscapes vary significantly across different industrial sectors, from energy and water to transportation and manufacturing. The INCD's role is to facilitate this exchange, ensuring that policies and guidance are informed by real-world operational challenges and that industry best practices are integrated into national cybersecurity strategies.

The INCD's emphasis on operational resilience is a key differentiator. This concept goes beyond traditional cybersecurity metrics and focuses on the ability of an organization or system to continue functioning, even when subjected to disruptive events, including cyberattacks. For OT systems, this means ensuring that essential services can be maintained, even if certain components are compromised or unavailable. This might involve developing redundant systems, implementing manual override capabilities, and ensuring that critical operational parameters can be monitored and controlled even in degraded modes. The INCD actively promotes this ethos, encouraging organizations to think about their OT security not just in terms of preventing unauthorized access but also in terms of maintaining operational continuity under duress. This is a vital distinction for critical infrastructure, where the consequences of service disruption can extend far beyond economic losses, impacting public safety and national security.

One of the notable aspects of Israel's approach, as exemplified by the INCD, is its perceived departure from the more alarmist rhetoric that is common in some cybersecurity discussions. While the potential for devastating cyber-attacks is a legitimate concern, the INCD's communication and policy initiatives tend to focus on building capabilities, fostering collaboration, and implementing practical solutions rather than solely on amplifying worst-case scenarios. This can be seen as a more sustainable and effective long-term strategy. By concentrating on tangible improvements and empowering stakeholders with the knowledge and tools to manage risks, the INCD aims to build a more robust and resilient national cybersecurity posture. This less alarmist, more solution-oriented approach can foster greater trust and cooperation between government agencies and the private sector, which is essential for effective OT security. It is a posture that, as we will see, finds a close parallel in the United Kingdom's National Cyber Security Centre, discussed in detail later in this chapter; the two nations arrive at a similar restraint from very different starting points.

The centralization under the INCD also facilitates a more proactive and forward-looking approach to OT security. By having a dedicated national body responsible for cyber defense, Israel can more effectively identify emerging threats, anticipate future technological challenges, and develop long-term strategies for adapting its defenses. This is particularly important in the OT domain, where the pace of technological change, the increasing convergence of IT and OT, and the growing adoption of Industrial Internet of Things (IIoT) devices present a constantly evolving threat landscape. The INCD can act as a central hub for intelligence gathering, threat analysis, and strategic planning, ensuring that Israel remains at the forefront of cyber defense innovation and preparedness. This includes investing in research and development, promoting the adoption of new security technologies, and fostering a skilled cybersecurity workforce capable of addressing the unique challenges of OT environments.

The INCD's effectiveness is also bolstered by the strong political will and government support it enjoys. The Israeli government has consistently recognized cybersecurity as a critical national security issue, and this recognition has translated into substantial investment and a clear mandate for the INCD. This high-level commitment is crucial for driving the necessary policy changes, resource allocations, and inter-agency coordination required to build a comprehensive national cybersecurity strategy. In the context of OT security, this political backing ensures that the protection of critical infrastructure is given the priority it deserves, enabling the INCD to enforce regulations, promote best practices, and drive the adoption of necessary security measures across all relevant sectors. This governmental endorsement also lends significant weight to the INCD's recommendations and directives, increasing their likelihood of being implemented by both public and private entities.

The Israeli model, with its emphasis on centralization, pragmatism, and resilience, offers several potential lessons for other nations, including the United States, as they grapple with the complex challenge of securing their OT systems. While the specific political, economic, and technological contexts of each nation are unique, the core principles of Israel's approach may hold broader applicability. The benefits of a unified national cyber authority, capable of coordinating efforts across government and industry, are significant. Such a body can ensure a more consistent and effective approach to policy development, incident response, and threat intelligence sharing. The pragmatic focus on tangible risks and operational resilience, as opposed to an over-reliance on fear-based rhetoric, can lead to more efficient resource allocation and a more sustainable security posture. Furthermore, the emphasis on public-private partnerships and information sharing is crucial for fostering a collaborative ecosystem where all stakeholders play an active role in safeguarding critical infrastructure.

In conclusion, Israel's National Cyber Directorate represents a significant effort to create a centralized, pragmatic, and resilience-focused approach to national cybersecurity, with a particular emphasis on the protection of critical infrastructure and its embedded OT systems. Its strategy of consolidating responsibilities, prioritizing tangible risks, fostering collaboration, and emphasizing operational continuity provides a valuable framework for understanding how a nation can effectively address the evolving cyber threats facing its industrial base. The INCD's success lies not in eradicating threats, which would be an unrealistic aspiration, but in building a robust and adaptive defense ecosystem that can withstand, respond to, and recover from cyber incidents with minimal disruption. This pragmatic, coordinated, and resilience-oriented model offers a compelling alternative to more fragmented or fear-driven approaches, demonstrating the potential for effective OT security through strategic leadership and practical implementation. The INCD's continuous adaptation and evolution in response to the dynamic threat landscape underscore the importance of agility and foresight in national cybersecurity efforts, particularly in the critical domain of operational technology. A comparable restraint, expressed through a very different institutional design, animates the United Kingdom's approach, to which this chapter returns in its dedicated treatment of the National Cyber Security Centre below.

The European Union: Harmonization and Regulation

The European Union presents a compelling case study in the pursuit of harmonized cybersecurity for Operational Technology (OT). Unlike the more nation-state-centric approaches seen elsewhere, the EU's strategy is inherently multilateral, built upon a foundation of shared principles and a desire to foster a cohesive digital single market. This ambition, while laudable, introduces a layer of complexity as it seeks to reconcile diverse national infrastructures, legal traditions, and threat perceptions under a common regulatory umbrella. At the heart of the EU's efforts lies a significant focus on establishing a baseline level of cybersecurity resilience across critical sectors, recognizing that a weakness in one member state can have cascading effects across the Union.

A pivotal element of this strategy is the Network and Information Security (NIS) Directive and its successor, NIS2. The NIS framework was groundbreaking in its ambition to establish a consistent level of cybersecurity across the EU. For OT, this meant extending cybersecurity requirements beyond traditional IT domains into sectors that underpin the functioning of society and the economy. The directive mandated that operators of essential services (OES), spanning energy, transport, banking, financial market infrastructures, health, drinking water, and digital infrastructure, implement appropriate security measures to manage risks and prevent incidents. This was not merely a suggestion; it imposed obligations, including security requirements, incident reporting, and oversight by competent authorities. The NIS Directive's initial focus on essential services provided a crucial entry point for addressing OT security, acknowledging that the physical processes controlled by these systems are fundamental to daily life and economic stability. It compelled organizations to move beyond ad-hoc security practices and adopt a more structured, risk-based approach to safeguarding their OT environments. The emphasis was on ensuring business continuity and service integrity, which required robust OT security.

The NIS2 Directive, adopted in late 2022 (Directive (EU) 2022/2555) with a transposition deadline of 17 October 2024, was enacted to address the shortcomings of NIS1 and the evolving threat landscape. It strengthens the original requirements and expands their scope. It broadens the definition of essential entities, encompassing more sectors and sub-sectors, many of which have significant OT footprints. This expansion means that a wider array of organizations operating critical infrastructure and supply chains are now subject to stricter cybersecurity obligations. NIS2 also introduces more specific security measures, such as enhanced incident reporting obligations, stricter supply chain security requirements, and a greater emphasis on risk management. For OT environments, this translates into a demand for greater visibility into the security posture of their entire ecosystem, including third-party suppliers and the devices that form part of their industrial control systems. The directive's push for more granular security measures, such as incident response planning, business continuity, and the use of cryptography, directly addresses many of the vulnerabilities inherent in OT systems. It pushes organizations to think holistically about their security, encompassing technical controls, organizational processes, and human factors, all within the context of their industrial operations.

Integral to the EU's regulatory architecture is the European Union Agency for Cybersecurity (ENISA). ENISA plays a crucial, multifaceted role in supporting the implementation of EU cybersecurity policy, including that pertaining to OT. It acts as a center of expertise, providing guidance, recommendations, and technical assistance to member states and EU institutions. For OT security, ENISA's contribution is invaluable. The agency has been instrumental in developing frameworks and guidelines that help organizations understand and implement the requirements of the NIS directives. This includes work on risk assessment methodologies tailored for critical infrastructure, guidance on incident reporting, and the development of best practices for securing industrial control systems. ENISA's research and reports often shed light on emerging threats and vulnerabilities specific to OT, providing a vital intelligence function that informs policy and industry action. Furthermore, ENISA facilitates cooperation and information sharing among member states, fostering a collective approach to cybersecurity challenges. This collaborative spirit is essential for addressing a threat landscape that transcends national borders and for developing a consistent understanding of OT security risks across the Union.

The EU's emphasis on harmonization is a defining characteristic of its approach. The goal is to create a level playing field across member states, ensuring that businesses operating in multiple EU countries do not face a patchwork of disparate and potentially conflicting cybersecurity regulations. This regulatory alignment aims to simplify compliance for multinational corporations and foster greater trust in the digital economy. For OT, this means that a company operating a power grid in Germany should, in principle, be working towards similar security objectives and implementing comparable controls as a company managing a water treatment facility in Spain. This harmonization is intended to reduce the risk of regulatory arbitrage and to build a stronger, more resilient cybersecurity posture across the entire Union. While achieving perfect uniformity is an ongoing challenge, the EU's persistent drive towards harmonized standards creates a more predictable and robust environment for OT security investment and implementation. It pushes national regulators to adopt consistent interpretations of directives and encourages the development of common technical standards and certification schemes.

Data protection, often seen as a distinct area of regulation, is also deeply intertwined with OT security within the EU framework, particularly through the General Data Protection Regulation (GDPR). While GDPR primarily concerns personal data, its principles of security by design and by default, as well as its requirements for data breach notification, have indirect but significant implications for OT. Many OT systems may process or be connected to systems that handle personal data, either directly (e.g., employee access logs) or indirectly (e.g., data linked to service delivery). A data breach stemming from a compromised OT system could therefore trigger GDPR obligations, underscoring the need for robust security measures within these industrial environments. This integration of data protection principles encourages a more comprehensive approach to security, where the protection of sensitive information is considered alongside the operational integrity of industrial processes. It elevates the importance of security governance and incident response planning within organizations, ensuring that OT security is not an isolated technical concern but a part of a broader data governance and risk management strategy.

Cross-border cooperation is another cornerstone of the EU's OT security strategy. Recognizing that cyber threats do not respect national boundaries, the EU has actively promoted mechanisms for collaboration among member states. This includes initiatives like the Cooperation Group and CSIRTs Network established under the NIS Directive, which facilitates the exchange of information, best practices, and coordinated responses to cybersecurity incidents. For OT, this cooperation is particularly vital. Sharing threat intelligence about attacks targeting industrial control systems, for instance, can enable other member states to preemptively strengthen their defenses. Similarly, joint exercises and simulations involving multiple member states can test and improve coordinated incident response capabilities across borders, which is crucial for critical infrastructure that often spans or interconnects with neighboring countries. ENISA plays a key role in facilitating this cooperation, serving as a neutral platform for dialogue and joint action. This collaborative framework is designed to build collective resilience, moving beyond individual national efforts to a more unified and effective Union-wide defense.

The EU's multinational approach, while inherently more complex due to the need for consensus-building and the integration of diverse national systems, offers distinct advantages. It has the potential to foster a more consistent and less fear-driven security posture across member states. By setting common standards and promoting shared understanding, the EU aims to avoid the fragmentation that can arise from solely national approaches. This can lead to more predictable regulatory environments, greater investment in cybersecurity technologies and expertise, and ultimately, a more resilient

EU-wide critical infrastructure. The emphasis on harmonization and cooperation encourages a long-term strategic vision for OT security, rather than reactive responses driven by individual incidents or immediate national anxieties. This structured, multilateral approach, championed through directives like NIS and supported by agencies like ENISA, represents a significant effort to collectively secure the digital and physical underpinnings of the European Union's economy and society, building a foundation of trust and resilience in an increasingly interconnected world.

The consistent application of these frameworks across a large bloc of nations inherently promotes a higher baseline of security. Instead of isolated pockets of excellence or significant vulnerabilities, the aim is to lift the security posture of all member states, recognizing that interdependence means a single weak link can compromise the entire chain. This collective ambition is what truly sets the EU's OT security framework apart, striving for a synchronized and robust defense against the pervasive and evolving cyber threats that target the very fabric of modern life.

ENISA's role extends beyond simply providing guidance; it actively works to build a common understanding and a shared vocabulary around OT cybersecurity. Through its publications, workshops, and expert groups, ENISA helps to bridge the gap between IT and OT security professionals, fostering a culture of collaboration within organizations. It promotes the adoption of international standards and best practices where appropriate, while also recognizing the unique characteristics of OT that may require tailored solutions. This includes addressing issues such as the long lifecycle of industrial equipment, the difficulty of patching operational systems without disrupting production, and the critical need for real-time performance and availability. ENISA's work in areas like the development of sector-specific cybersecurity guidelines for critical infrastructure provides practical tools and frameworks that organizations can adapt to their specific operational contexts. This granular, sector-aware approach is vital for ensuring that cybersecurity measures are effective and proportionate within the diverse landscape of European industry.

The EU's regulatory framework, therefore, aims to create a resilient and secure digital single market where businesses and citizens can operate with confidence. For OT, this translates into a concerted effort to elevate cybersecurity from a niche technical concern to a strategic imperative for national security and economic prosperity. The harmonization of regulations, the emphasis on risk management and incident reporting, the focus on supply chain security, and the continuous support from ENISA all contribute to building a more robust and consistent defense against cyber threats targeting industrial control systems and critical infrastructure across the Union. This collaborative, standards-driven, and increasingly comprehensive approach seeks to ensure that the benefits of digitalization in industrial sectors are not undermined by preventable cybersecurity failures, thereby fostering a more secure and prosperous European future.

Common Themes and Contrasting Philosophies

Securing Operational Technology (OT) is a global challenge with diverse national responses. By examining the approaches taken by other nations, we can glean invaluable lessons, identifying common threads of success and understanding the nuances that make certain strategies more effective than others. This comparative analysis is crucial for refining our own policies and strengthening our resilience against the escalating threats to critical infrastructure. While the United States has certainly made strides, a deeper dive into international experiences can offer a fresh perspective, highlighting potential blind spots and innovative solutions. One of the most striking observations from international comparisons is the varying degrees of emphasis placed on centralized coordination versus decentralized autonomy. Some nations have opted for a top-down, highly centralized model, often driven by national security imperatives. This approach typically involves a dedicated government agency with broad authority to set standards, conduct assessments, and even direct remediation efforts for critical infrastructure. The perceived advantage is a unified strategy, ensuring common goals and efficient resource allocation based on a national threat assessment. This can lead to standardized security controls and a clear chain of command during a major incident. However, such centralization can sometimes be rigid, struggling to adapt to the unique operational realities and diverse technological landscapes of individual sectors or organizations. The sheer complexity of modern industrial control systems and their inherent differences across industries can make a one-size-fits-all mandate difficult to implement effectively.

Conversely, other nations have leaned towards a more decentralized or collaborative model, emphasizing public-private partnerships and industry-led initiatives. In these frameworks, the government often plays a role of facilitator, providing guidance, sharing threat intelligence, and incentivizing security improvements, but leaving the primary responsibility for implementation and risk management to the private sector operators of critical infrastructure. This approach can foster greater agility and innovation, as industry experts are often best positioned to understand the specific risks and develop tailored solutions for their own systems. It also respects the proprietary nature of many industrial processes and encourages a sense of ownership and accountability within the industry itself. The challenge here lies in ensuring a sufficient baseline of security across all operators, preventing a "race to the bottom" where cost savings might compromise essential security measures, and ensuring effective information sharing between government and industry. The United States' own approach has historically seen elements of both centralization and decentralization, with various agencies having responsibilities, but a cohesive, overarching OT cybersecurity strategy has been a longstanding challenge.

A critical factor in the success of any OT security strategy is the quality and nature of its threat assessment. Nations that have achieved notable progress often demonstrate a commitment to evidence-based threat intelligence. This means moving beyond generic risk assessments to a dynamic understanding of the specific adversaries, their tactics, techniques, and procedures (TTPs) targeting industrial control systems. Such an approach requires robust mechanisms for collecting, analyzing, and disseminating actionable threat information, often involving close collaboration between intelligence agencies, law enforcement, and the private sector. When threat assessments are granular and relevant, they enable the prioritization of resources and the development of more effective defensive measures. For instance, understanding that a particular threat actor is actively targeting the energy sector's Supervisory Control and Data Acquisition (SCADA) systems with specific malware allows for the deployment of targeted defenses and enhanced monitoring within that sector. Without this clear understanding of the adversary, security efforts can become unfocused and less impactful.

The effectiveness of public-private partnerships (PPPs) in OT security also varies significantly across different national contexts. In countries where PPPs are well-established and deeply integrated, there is often a strong foundation of trust and shared understanding between government entities and private sector operators. These successful partnerships are characterized by open communication channels, mutual respect for each other's expertise and constraints, and clear delineation of roles and responsibilities. The government might provide classified threat intelligence and regulatory frameworks, while industry partners offer operational insights, share incident data (often anonymously), and implement security controls. This symbiotic relationship allows for a more comprehensive and adaptive approach to cybersecurity than either sector could achieve alone. Examples of this include information sharing and analysis centers (ISACs) that are effectively resourced and supported, or collaborative initiatives focused on developing sector-specific security standards and best practices.

In contrast, less effective PPPs may suffer from a lack of trust, information silos, or a perception of overly burdensome regulation from the government's side, or a reluctance from industry to share sensitive information. Where the government is seen as purely a regulator seeking to impose costs, rather than a partner in ensuring collective security, engagement can be limited. Similarly, if industry perceives government efforts as not aligned with their operational realities or as merely a compliance exercise, the partnership will falter. The U.S. has numerous ISACs and sector-specific agencies like CISA that strive to foster these partnerships, but the depth and breadth of their success can be uneven across different critical infrastructure sectors, influenced by the unique characteristics and maturity of each.

Examining successful international models offers several key takeaways. Firstly, the importance of a clear national strategy that defines roles, responsibilities, and objectives for OT cybersecurity cannot be overstated. Whether this strategy is implemented through centralized mandates or collaborative frameworks, a cohesive vision is essential. Secondly, a strong emphasis on evidence-based threat intelligence and a proactive understanding of adversary behavior is paramount. This requires investment in intelligence gathering, analysis, and dissemination capabilities, as well as mechanisms to effectively share this intelligence with critical infrastructure operators. Thirdly, well-structured and trusted public-private partnerships are a powerful force multiplier. These partnerships must be built on a foundation of mutual understanding, shared goals, and clear benefits for all involved, moving beyond a purely compliance-driven relationship.

Furthermore, the concept of "security by design" and "security by default" is gaining traction internationally, particularly in discussions surrounding new OT deployments or significant upgrades. This proactive approach aims to integrate cybersecurity considerations from the very inception of a system's lifecycle, rather than attempting to bolt on security measures later. This is a crucial shift, acknowledging that many legacy OT systems were designed with operational reliability and safety as the sole priorities, often with little regard for cybersecurity. International efforts are increasingly focusing on promoting these principles through standards development, procurement guidelines, and even regulatory requirements, pushing manufacturers and operators to build more secure systems from the ground up. This is a long-term endeavor, but one that holds immense promise for building a more resilient future.

The role of international standards and certifications in promoting OT security is also a significant area of development. Many nations are aligning their national frameworks with internationally recognized standards, such as those from the International Electrotechnical Commission (IEC) or the International Organization for Standardization (ISO). This not only facilitates interoperability and trade but also provides a common language and a benchmark for security practices. The development and adoption of sector-specific cybersecurity standards for industrial control systems, tailored to the unique requirements of different industries, are also proving to be highly effective. These standards provide clear guidance on what constitutes adequate security measures, helping organizations to assess their own posture and to select appropriate technologies and processes.

The approach taken by some European nations, for instance, highlights the value of a multilayered strategy that combines regulatory mandates with strong industry collaboration and a focus on specific vulnerabilities. While the EU's NIS2 directive provides a broad regulatory framework, the implementation often involves national-level initiatives that are tailored to local contexts. These often include dedicated government bodies focused on critical infrastructure protection, national incident response teams with specialized OT expertise, and government-supported programs for industry training and awareness. The emphasis on information sharing and the development of common incident response protocols across member states also demonstrates a recognition that cyber threats do not respect borders, and that collective defense is more effective. Ultimately, the lessons from other nations converge on a few core principles: proactive threat intelligence, strong public-private collaboration, a risk-based approach tailored to OT realities, a focus on fundamental security hygiene, and the development of a skilled workforce. While the specific implementation details may vary, the underlying commitment to understanding the unique challenges of OT security and developing adaptable, resilient defenses is a universal requirement for safeguarding critical infrastructure in the face of evolving cyber threats. The U.S. can and should learn from these diverse international experiences to refine its own policies, foster greater collaboration, and ultimately enhance the security and resilience of its vital industrial systems. The challenge is not to replicate any single model, but to synthesize the most effective elements into a robust and adaptable framework that can stand the test of time and evolving threats.

The United Kingdom: Demystifying Security Through Clarity — Contrasting Threat Perceptions

The United Kingdom's approach to securing its Operational Technology (OT) environment, particularly through the National Cyber Security Centre (NCSC), offers a distinctive perspective on national cybersecurity strategy, and a particularly instructive one for the central argument of this book. Where Israel arrives at restraint through centralized, security-state pragmatism and the EU arrives at it through harmonized regulation, the UK arrives at it through something rarer in this field: an explicit institutional decision to *demystify* the threat rather than to amplify it. While many nations grapple with the complexities of OT security by establishing dedicated agencies or heavily emphasizing regulatory enforcement, the NCSC has carved out a niche by focusing on clarity, collaboration, and the fundamental principles of good cyber hygiene. This strategy aims to "demystify" cybersecurity, making it accessible and actionable for a broad range of organizations, including those operating critical national infrastructure. Unlike some national narratives that can amplify a sense of existential threat, the NCSC's communication style often leans towards empowering organizations with practical knowledge and tools, thereby fostering a more proactive and less fear-driven security posture.

The NCSC, established in October 2016 as part of GCHQ, operates with a mandate to provide unified cyber security leadership across the UK. Its remit is broad, encompassing the protection of critical national infrastructure, government systems, and the wider public from cyber threats. In the realm of OT security, this translates into a significant effort to bridge the often-perceived gap between IT and OT security practices. That it sits inside GCHQ, the United Kingdom's signals-intelligence agency, makes its public posture all the more notable: this is not a civilian outreach office insulated from the hard edge of nation-state threat intelligence, but the defensive arm of an organization that sees genuine adversary traffic up close, and which has, in the view of many observers, tended to favor calm explanation over alarm in its public communications.

A cornerstone of the NCSC's strategy is its commitment to providing clear, actionable, and authoritative guidance. This is evident in its extensive library of advisories, best practice guides, and technical advice, many of which are directly applicable to OT environments or can be adapted by OT professionals. The NCSC eschews overly technical jargon where possible, aiming to communicate complex security concepts in a manner that is understandable to a wide audience, from IT security managers to operational engineers and even executive leadership. For OT security, this means providing guidance on topics such as network segmentation, secure remote access, vulnerability management for industrial control systems (ICS), and the importance of asset inventory, all presented in a way that emphasizes practical implementation rather than abstract theory.

This emphasis on clarity and actionability serves a critical purpose in demystifying cybersecurity. Many organizations, particularly those with legacy OT systems and a workforce more accustomed to operational processes than digital threats, can find the cybersecurity landscape overwhelming. The NCSC's approach aims to break down these barriers by offering a structured path to improving security. Their guidance aimed at smaller organisations (e.g., the Small Business Guide) and the Cyber Essentials certification scheme (which, although widely adopted by SMEs, is open to organisations of any size) for instance, can be scaled and adapted for larger industrial enterprises, focusing on foundational controls that yield significant security benefits. This methodical approach helps reduce the panic that can arise from perceived omnipresent threats, enabling organizations to focus on tangible improvements.

It is in the assessment framework that underpins the United Kingdom's NIS regulatory regime, however, that the deepest philosophical difference reveals itself. The NCSC's Cyber Assessment Framework (CAF) is not a compliance checklist in the conventional sense, and the distinction is not cosmetic. Where a checklist regime asks an operator to confirm that it has installed a defined list of controls, ticking boxes that can be satisfied on paper while the underlying system remains brittle, the CAF is built around *outcomes*. It is organized into a set of high-level security and resilience objectives, expressed as the results an organization is expected to achieve rather than the specific products it is expected to buy. An operator is asked not "have you deployed control X?" but "can you demonstrate that you understand, manage, and can recover the systems on which essential services depend?" The framework's contributing outcomes and indicators of good practice are designed to be assessed on a graduated basis, recognizing that a small water utility and a national transmission operator face very different risk profiles and that proportionality, not uniformity, is the goal. This outcome-based philosophy matters enormously for the argument of this book. A compliance-checklist culture is precisely the soil in which the fear machine grows best: when security is defined as the acquisition of a list of products, every new fear becomes a new line item, and the vendor's incentive is to lengthen the list indefinitely. An outcomes regime, by contrast, asks whether the essential service can actually survive a disruption, a question no product can answer on its own, and one that tends to redirect attention away from the purchase of reassurance and toward the harder, less profitable work of genuine resilience. The CAF thereby encodes, at the level of national regulation, the very skepticism toward fear-driven procurement that runs through this book.

This regulatory architecture sits underneath, and gives concrete force to, the NCSC's broader posture. The Cyber Assessment Framework functions in practice as the assessment basis against which operators of essential services in the United Kingdom can be measured, the domestic counterpart to the kind of obligations the European Union pursues through its NIS framework, but distinguished by its insistence on judging the outcome achieved rather than the inputs assembled. For OT environments in particular, this is a meaningful difference. The systems that run a power station or a water-treatment works were built for reliability and longevity, not for the tidy enumeration of security controls; an outcome-based assessment can accommodate the messy reality of a twenty-year-old programmable logic controller that cannot simply be patched on a vendor's schedule, asking instead how the operator manages, monitors, and contains the risk that such a device represents. A checklist, applied mechanically to the same device, would either fail it outright or invite the operator to claim a compliance that does not reflect operational reality. The UK's framework is, in this sense, built for the world OT security actually inhabits.

Furthermore, the NCSC's strategy is deeply rooted in fostering collaboration and information sharing, particularly with industry. Recognizing that the government cannot secure critical infrastructure alone, the NCSC actively engages with sector-specific bodies, industry forums, and individual organizations. This collaboration is vital for OT security, as the operational contexts and risk profiles of different industrial sectors vary considerably. The NCSC acts as a central hub, facilitating the flow of threat intelligence, best practices, and incident lessons learned. This is achieved through various channels, including dedicated liaison officers, industry roundtables, and information-sharing partnerships. For OT, this means that insights gained from an incident in the energy sector, for example, can be quickly disseminated to other critical infrastructure operators, allowing them to proactively bolster their defenses.

This collaborative model also extends to practical exercises and simulations. The NCSC frequently conducts cyber exercises, often in partnership with industry, to test and improve response capabilities. These exercises are designed to simulate realistic cyberattacks on critical infrastructure, allowing organizations to practice their incident response plans, test their technical defenses, and refine their coordination mechanisms. Such events are invaluable for OT environments, as they provide a safe space to experience the challenges of responding to cyber incidents that have a direct impact on physical processes, helping to build confidence and competence among operational staff.

The NCSC's emphasis on "practical cyber hygiene" is another key differentiator. Rather than solely focusing on the latest, most sophisticated attack vectors, the NCSC consistently highlights the importance of fundamental security measures. This includes basic practices such as robust password policies, regular software patching, network segmentation, user awareness training, and maintaining up-to-date asset inventories. For OT systems, these foundational elements are often overlooked or are more challenging to implement due to the nature of the systems. The NCSC's persistent advocacy for these measures, coupled with practical guidance on how to implement them in OT contexts, helps to raise the baseline security posture across critical national infrastructure. They understand that many successful cyberattacks on OT exploit these basic vulnerabilities, and by addressing them, organizations can significantly reduce their risk exposure.

This approach contrasts with a narrative that might solely focus on the capabilities of advanced persistent threats (APTs) or nation-state actors. While the NCSC certainly monitors and advises on these high-level threats, its public-facing strategy often prioritizes empowering organizations to defend against the most common and impactful attack methods. This is particularly relevant for OT, where the consequences of even a seemingly unsophisticated attack can be severe. By providing clear guidance on cyber hygiene, the NCSC aims to equip organizations with the resilience to withstand a wider range of threats, thereby reducing the overall risk to critical national infrastructure.

The NCSC's efforts to demystify cybersecurity and reduce unnecessary panic are also reflected in its communication strategy. While the seriousness of cyber threats is never downplayed, the NCSC strives for a balanced and objective tone. This is achieved by providing evidence-based advice, clearly explaining the nature of threats, and offering concrete steps for mitigation. This measured approach helps to foster a more rational and effective response to cybersecurity challenges. Instead of fostering a sense of helplessness or constant alarm, the NCSC encourages a mindset of proactive risk management and continuous improvement. This is crucial for OT security, where an overly alarmist tone could lead to misallocation of resources or the implementation of overly burdensome controls that could disrupt operations.

This measured institutional voice is not merely an accident of tone; it reflects a culture of official skepticism toward threat-inflation that is comparatively rare among national cyber agencies, and it has on at least one occasion been stated with remarkable bluntness. In February 2017, the NCSC's then-Technical Director, Ian Levy, publicly compared a large part of the threat-intelligence industry to "medieval witchcraft," arguing that vendors were dressing up ordinary criminal activity as the work of unstoppable adversaries in order to sell protection against it.[1] That a sitting, senior official of the country's own cyber-defense establishment would say such a thing in public, rather than a disgruntled customer or an outside academic, tells us something important about the official culture from which the NCSC's guidance flows: it is a culture prepared to call exaggeration by its name. The episode is examined at length in Chapter 4, where it anchors the discussion of the vendor ecosystem; here it is enough to note that the same skepticism which produced Levy's "magic amulet" remark is the skepticism that produces the NCSC's calm advisories and its outcome-based regulatory framework. The tone and the policy are of a piece.

Moreover, the NCSC's role as a central, authoritative source of advice builds trust. By being consistently reliable and transparent, the NCSC has established itself as a trusted partner for both government and industry. This trust is essential for encouraging the open sharing of sensitive information about vulnerabilities and incidents, which is crucial for effective OT security. When organizations feel confident that the NCSC will handle information responsibly and provide valuable, actionable advice, they are more likely to engage and collaborate. It is worth dwelling on the reciprocal nature of this dynamic, because it is the mechanism by which a less alarmist posture actually produces better security rather than merely a calmer mood. An agency that cries wolf, that frames every advisory in the language of imminent catastrophe, eventually trains its audience to discount its warnings; the operators it most needs to reach learn to tune out the noise. An agency that reserves alarm for the rare occasions that genuinely warrant it, and otherwise communicates in measured, evidence-based terms, retains its credibility, and credibility is the currency in which voluntary information-sharing is paid. The NCSC's restraint, in other words, is not softness; it is a deliberate investment in the signal-to-noise ratio of national cyber communication.

The NCSC's strategy also involves a strong focus on research and development, ensuring that its advice and guidance are informed by the latest understanding of cyber threats and technologies. This includes research into emerging threats targeting OT systems, such as those related to the Industrial Internet of Things (IIoT), cloud integration, and advanced automation. By staying at the forefront of technological developments and threat landscapes, the NCSC can provide timely and relevant advice to help organizations prepare for future challenges in the OT domain.

In essence, the NCSC's approach to OT security can be characterized by its pragmatism, clarity, and collaborative spirit. It recognizes that effective cybersecurity is not achieved through a single silver bullet or a purely technical solution, but through a combination of clear guidance, robust fundamental practices, and strong partnerships. By demystifying cybersecurity, promoting practical cyber hygiene, and fostering a collaborative ecosystem, the NCSC empowers UK organizations, including those responsible for critical national infrastructure, to build and maintain a resilient defense against the evolving threat landscape. This deliberate strategy aims to foster a sustainable and effective approach to OT security, moving beyond fear and towards empowered action.

The United Kingdom's example therefore opens onto a larger point, one that the remainder of this chapter develops by stepping back from any single national agency to the contrasting *perceptions* of threat that shape policy across borders. The global landscape of Operational Technology (OT) security is far from monolithic. While the urgency and scale of cyber threats to critical infrastructure are universally acknowledged, the *perception* of these threats, and consequently the policy responses they engender, diverge significantly across national borders. This divergence is not merely a matter of differing threat landscapes or resource availability; it often reflects fundamental differences in national security philosophies, historical experiences, and societal risk tolerances. Examining these contrasting threat perceptions is crucial for understanding why certain OT security strategies gain traction in one nation while others falter, and it offers a potent counterpoint to the often alarmist narratives that can shape policy in the United States.

In many countries, particularly those with robust democratic traditions and a strong emphasis on governance and rule of law, the approach to OT security tends to be more deliberate and rooted in established institutional frameworks. The United Kingdom, as the preceding discussion of the NCSC and its Cyber Assessment Framework illustrates, is a paradigmatic case of exactly this institutional deliberateness. These nations, even those facing significant geopolitical pressures or direct cyber campaigns, often prioritize building resilient systems and ensuring continuity of operations through thoughtful policy development and collaboration. The emphasis is frequently on creating an environment where critical infrastructure operators can effectively manage risks based on a clear understanding of their specific vulnerabilities and the evolving threat landscape, rather than being driven by a constant influx of worst-case scenario pronouncements. This can lead to policies that are more sustainable, less prone to knee-jerk reactions, and better integrated into the broader economic and societal fabric. The drive is not necessarily to eliminate all risk, an impossible and ultimately counterproductive goal, but to ensure that critical services can withstand and recover from disruptions, maintaining public trust and essential functions.

This more measured approach is often evident in how these nations frame the debate around OT security. Instead of focusing predominantly on the catastrophic potential of a single, devastating cyberattack, the discourse tends to revolve around the ongoing, systemic challenges of maintaining security and resilience in complex, interconnected industrial environments. This involves a deep understanding of the operational realities of sectors like energy, water, transportation, and manufacturing, recognizing that these systems were often designed for reliability and safety, with cybersecurity as a secondary, or even absent, consideration. Consequently, policy initiatives frequently target long-term solutions, such as fostering a culture of security within organizations, promoting best practices for system design and maintenance, and building robust incident response capabilities that are regularly tested and refined. The narrative is one of continuous improvement and adaptation, rather than a perpetual state of emergency. It is no coincidence that this is precisely the language in which the NCSC's outcome-based framework is written: a framework that asks whether a service can recover is, by its very construction, a framework organized around systemic resilience rather than around the spectre of a single annihilating blow.

A key driver for this less alarmist posture in many international contexts is a mature understanding of public-private partnerships. In countries where such collaborations are deeply embedded, the government often acts as a facilitator and enabler, rather than a command-and-control entity. This partnership model is built on a foundation of trust, where critical infrastructure operators share information about threats and vulnerabilities, and in return, receive actionable intelligence, regulatory clarity, and support for security investments. The focus is on shared responsibility and mutual benefit. The government's role is to set the overarching policy framework, provide essential resources (like threat intelligence derived from national security efforts), and ensure a baseline level of security, while industry retains the primary responsibility for implementing and managing security measures tailored to its specific operational context. This contrasts sharply with approaches that might be perceived as overly prescriptive or burdensome, potentially stifling innovation and creating resistance within the private sector.

Furthermore, the international community has witnessed the evolution of OT security not as a purely technical problem, but as a complex socio-technical challenge deeply intertwined with national resilience and economic stability. This holistic perspective encourages a focus on outcomes, the continued functioning of essential services, rather than solely on inputs, such as the number of security controls implemented or compliance with a specific set of regulations. When a nation's policy discourse centers on resilience, it naturally leads to strategies that emphasize redundancy, contingency planning, rapid recovery, and the ability to adapt to unforeseen circumstances. This requires a sophisticated understanding of critical infrastructure interdependencies and the cascading effects that an incident in one sector could have on others. The aim is to build systems that are not just secure, but inherently robust and capable of absorbing shocks. The reader will recognize in this input-versus-outcome distinction the same axis along which the United Kingdom's Cyber Assessment Framework was earlier shown to depart from a conventional compliance checklist; what the CAF expresses as a matter of national regulation, the broader international turn toward resilience expresses as a matter of policy philosophy.

The contrast with the United States' approach to OT security is often stark. In the U.S., the narrative surrounding OT cybersecurity has, at times, been heavily influenced by a potent combination of national security concerns, media amplification, and a perceived governmental inclination towards highlighting potential existential threats. This has led to a policy environment that can sometimes be characterized by a sense of perpetual urgency and an emphasis on the most dire, albeit often hypothetical, attack scenarios. While such warnings can galvanize action, they can also create an environment of heightened fear, potentially leading to policy decisions that are reactive, disproportionate, or less sustainable in the long run. The "fear machine," as it might be termed, can overshadow the more pragmatic, incremental, and collaborative efforts that are often the bedrock of true, long-term resilience. The juxtaposition is sharpened considerably when one recalls that one of the most quotable official indictments of that very fear machine, Ian Levy's "medieval witchcraft," came not from an American critic of the American posture but from a senior official of a close ally's defensive establishment, looking at the same global industry and refusing to play along.

This American tendency towards a more alarmist framing can also impact the dynamics of public-private interaction. When the government's communication is perceived as overly focused on apocalyptic threats, it can breed a degree of skepticism or even resistance from industry partners who may feel their practical concerns and operational realities are being overlooked. The emphasis can shift from collaborative problem-solving to a compliance-driven relationship, where private sector entities feel they are being dictated to, rather than being engaged as essential partners in a shared endeavor. This can create an atmosphere where the perceived risks are amplified, but the actual development of practical, effective security measures is hindered by a lack of trust and open dialogue. The focus can become more on avoiding blame and meeting regulatory checkboxes than on fundamentally improving security posture. Here again the structural point made earlier about the Cyber Assessment Framework returns with added force: a regime built on checkboxes invites exactly this defensive, blame-avoiding posture, whereas a regime built on demonstrable outcomes gives the operator no box to hide behind and the regulator no list to wave, leaving both parties with the harder but more honest question of whether the essential service would in fact survive.

Moreover, the sheer volume of threat intelligence and advisories disseminated in the U.S. can, paradoxically, lead to an overload of information, making it difficult for operators to discern the most critical and actionable insights. While a comprehensive flow of information is generally desirable, an undifferentiated stream of warnings, many of which may be low-probability but high-impact, can dilute the focus on more immediate and probable threats. This contrasts with approaches in other nations where threat intelligence dissemination is often more curated, contextualized, and tailored to specific sectors or risk profiles, ensuring that operators receive information that is directly relevant to their operational environment and decision-making processes. The goal becomes not just to inform, but to empower informed action. This is the operational expression of the credibility argument made above in connection with the NCSC: an agency that floods its constituents with undifferentiated alarm spends down the very trust on which curated, actionable, voluntarily-shared intelligence depends.

The philosophical underpinnings of national security also play a significant role. Some nations view OT security primarily through the lens of ensuring the continued provision of essential services and maintaining public confidence, framing it as a matter of national economic and social stability. This perspective naturally leads to policies that prioritize resilience, redundancy, and the ability to recover quickly from disruptions. The threat is seen as an inherent part of operating in a digital world, and the response is geared towards managing and mitigating that risk over the long term. In contrast, the U.S. approach has, at times, been more heavily influenced by a national security paradigm that emphasizes the identification and neutralization of adversaries and the prevention of catastrophic attacks, often driven by intelligence assessments of state-sponsored threats. While crucial, this focus can sometimes lead to an overemphasis on offense and defense against specific actors, potentially at the expense of building broader, systemic resilience that can withstand a wider range of threats, including non-state actors, insider threats, and even accidental disruptions. The United Kingdom is an especially illuminating control case here, because it dismantles a tempting but lazy explanation for the difference. One might suppose that calmer nations are simply nations that see less threat, that the NCSC speaks softly because it has less to fear. But the NCSC is part of GCHQ and serves as the UK's national technical authority on cyber security; it sits at the center of one of the most capable signals-intelligence enterprises in the world and has no shortage of genuine, state-sponsored adversary activity in its field of view. Its restraint is therefore not the product of ignorance or of a benign threat environment. It is a chosen posture, adopted by people who see the real threat clearly and have nonetheless concluded that calm, outcome-focused, credibility-preserving communication serves national resilience better than perpetual alarm.

This difference in emphasis can manifest in the types of investments and initiatives that are prioritized. Nations focused on resilience may heavily invest in R&D for inherently secure industrial control systems, robust backup and recovery infrastructure, and comprehensive emergency preparedness and response training for operational personnel. The United States, while also investing in these areas, may dedicate more resources to intelligence gathering, offensive cyber capabilities to deter adversaries, and the development of sophisticated defensive tools aimed at detecting and eradicating specific threats. While both approaches have their merits, the resilience-focused model often fosters a more sustainable and adaptive security posture for critical infrastructure, acknowledging that the threat landscape is constantly evolving and that a purely adversarial approach may not always be sufficient. The pursuit of invulnerability is a chimera; the pursuit of resilience is a pragmatic strategy for navigating an uncertain future.

These contrasting threat perceptions have profound policy implications. When OT security is framed as an existential threat, policy often tends towards mandates, penalties, and a top-down enforcement model. This can lead to a sense of urgency but may also foster a compliance-focused culture rather than one of proactive risk management and continuous improvement. Conversely, when OT security is approached as a fundamental aspect of national resilience and economic stability, policy tends to favor incentives, collaborative frameworks, and capacity building. This approach encourages a deeper engagement from industry, fosters innovation, and ultimately leads to more robust and sustainable security outcomes. The United Kingdom's Cyber Assessment

Framework can be read as a deliberate attempt to occupy the better of these two poles by design: it carries the force of a national regulatory expectation, and so is not merely voluntary exhortation, yet by framing its demands as outcomes to be demonstrated rather than penalties to be feared, it pulls toward proactive risk management rather than defensive box-ticking. The lessons from international experiences suggest that a balanced approach, one that acknowledges the reality of threats without succumbing to undue alarmism, and that prioritizes collaboration and resilience, is likely to be the most effective in safeguarding critical infrastructure in the long term. The challenge for policymakers is to cultivate a policy environment that supports this more measured and resilient path, the path that the United Kingdom, through an agency that sees the threat plainly and still declines to sell fear, has done as much as any nation to chart.

Chapter 8

THE ILLUSION OF CONTROL: SYSTEMIC VULNERABILITIES UNMASKED

The prevailing narrative surrounding cybersecurity, particularly concerning critical infrastructure and Operational Technology (OT), often gravitates towards the specter of sophisticated, well-resourced adversaries. We conjure images of nation-state actors, shadowy cybercriminal syndicates, or even ideologically driven hacktivists wielding advanced tools and exhibiting an almost superhuman ability to breach defenses. The fear machine sells the superweapon: the bespoke cyber-physical payload, the zero-day that leaps an air gap, the unseen hand that flips a switch and plunges a region into darkness. It is a compelling story, and like most compelling stories it is mostly cinema. The argument of this chapter is deliberately deflationary. When we go looking for the vulnerabilities that actually get exploited, that actually destabilize operations, we do not find the exotic. We find the mundane. We find the default password that was never changed, the device that was never supposed to face the internet but does, the credential that protected a major regional fuel supply with no second factor behind it. The gap between the threats the fear machine markets and the weaknesses that matter in practice is the central diagnostic finding of this book, and it is the work of this chapter to draw that gap precisely.

This is not to dismiss the threat posed by capable adversaries; it is undeniable and warrants serious consideration. But a focus on the "who" and the "how" of highly advanced attacks can inadvertently obscure a more pervasive and foundational set of vulnerabilities inherent in the OT systems themselves. These are not vulnerabilities introduced by external actors, but rather deeply embedded weaknesses that have accumulated over decades, creating a broad, often unacknowledged, attack surface. The uncomfortable truth is that this surface is exploitable by a far wider spectrum of threats than the narrative admits, from the most sophisticated to the most rudimentary, and that the rudimentary end of that spectrum does the overwhelming share of the actual damage. Understanding these systemic vulnerabilities is paramount, because they represent the low-hanging fruit that can destabilize operations regardless of the attacker's sophistication or intent, and because the resources currently committed to chasing the superweapon are, in large measure, resources not spent fixing the boring problems that get organizations breached.

Before we catalog those weaknesses, it is worth pausing on a finding so at odds with the prevailing narrative that it is best delivered by a hostile witness. Robert M. Lee is the chief executive of Dragos, a leading vendor of industrial-control-system security, a company whose commercial interest lies squarely in selling OT defense. He is not a skeptic of the field; he is one of its most prominent practitioners. And it is Lee who has stated flatly, in 2023, that "there have been zero known ICS vulnerabilities ever leveraged in any ICS cyberattack."[1] Dragos's own research has gone further, arguing that industrial organizations "may worry too much about ICS vulnerabilities"[2]: that the great majority of disclosed ICS CVEs sit deep inside networks where an attacker would already need extensive access to reach them, that they are not exploitable for the catastrophic scenario the headlines describe, and that, as a class, they have not been used in real attacks. The real OT intrusions, in this telling, do not turn on a clever exploit of a software flaw at all. They turn on legitimate functionality used illegitimately and on stolen credentials, an attacker logging in rather than breaking in.

Sit with the source of that claim for a moment. It comes from inside the industry that profits from vulnerability anxiety. It is the equivalent of the locksmith telling you that burglars almost never pick locks, they walk through doors people left unlocked. When the party with the strongest commercial incentive to inflate the danger of ICS vulnerabilities is the one telling you those vulnerabilities have rarely been the decisive lever in an actual attack, the burden of proof shifts hard onto the narrative that demands you patch them at any operational cost. Much of the CVE-and-patch panic that drives

OT security spending, the breathless advisory, the dashboard glowing red with "critical" findings, the audit that treats an unpatched device as a breach-in-waiting, is built on a premise its own most credible insiders do not endorse. This does not mean vulnerabilities are irrelevant, and it does not mean patching is pointless; it means the rarity of vuln-driven OT attacks has been systematically mislabeled as imminence, and that the mislabeling has a price measured in misdirected budgets and unattended fundamentals.

The Paradox of the Failed-Safe System

The phrase that titles this chapter, the illusion of control, is usually deployed in one direction only: against defenders, to argue that their sense of security is false, that beneath the dashboards and the firewalls the system is fragile and the attacker holds the initiative. That direction is real, and we will honor it. But the illusion of control cuts both ways, and the most instructive single incident in the OT record demonstrates the second, neglected edge.

In 2017, an attack now known as Triton, or Trisis, struck a petrochemical plant in Saudi Arabia, targeting its Schneider Electric Triconex safety controllers. What made it singular was its target: not the ordinary process controllers, but the Safety Instrumented System, the dedicated, independent layer of engineering whose entire purpose is to bring a plant to a safe state when something goes wrong. This is the nightmare scenario the fear machine invokes, an adversary reaching not merely into operations but into the last-resort safety logic itself. And here is the detail that the cinematic framing almost always omits. The attack came to light because it inadvertently tripped the Safety Instrumented System, causing a process shutdown that prompted the investigation.[3] The engineered safety layer did exactly the job it was built to do. The intruders' fumbling against an unfamiliar safety controller did not produce catastrophe; it produced the controlled stoppage that the controller existed to produce, and in stopping the plant it announced the intrusion.

The lesson of Triton is the paradox at the heart of this chapter. Real industrial plants are not the brittle digital abstractions of the threat brief. They are physical systems wrapped in layers of protection that long predate cybersecurity and that do not depend on it: pressure-relief valves that open on physics alone, interlocks that refuse impossible commands, mechanical trips, redundant sensors, and safety systems engineered, certified, and tested to fail into safety rather than into disaster. A cyberattack does not trivially defeat these layers, because many of them are not software at all and cannot be reasoned with by a remote intruder. The fear machine commits two symmetrical errors at once. It overstates the attacker's control, imagining a degree of precise, deterministic command over a foreign industrial process that real intrusions rarely achieve. And it understates the defender's control, ignoring the deep, redundant, physics-anchored safeties that make the leap from "compromised a controller" to "killed people" far longer and far less certain than the narrative allows. Triton is held up, correctly, as a sobering escalation, an adversary did reach for the safety layer. But it is at least as much a story about a safety layer that worked. Both readings are true, and a narrative that can only hold the first one is not describing the world; it is describing a movie.

The Mundane Reality of What Actually Gets Exploited

If the superweapon is rare and the engineered safeties are real, what is it that actually goes wrong? The honest answer is unglamorous to the point of embarrassment, and that embarrassment is itself diagnostic, because boring problems do not sell conference keynotes, security appliances, or supplemental budget lines.

Consider the campaign attributed to the group calling itself Cyber Av3ngers across 2023 and 2024. The vulnerability they exploited was not a zero-day, not a bespoke implant, not a feat of cryptographic subversion. It was a Unitronics programmable logic controller exposed directly to the internet whose default password ("1111") had not been changed.[4] The result of breaching this supposedly critical infrastructure was a defaced screen. There was no loss of service, no poisoned water, no cascading regional failure, the catastrophe inventory the fear machine keeps in stock. An adversary reached an industrial device because the device was on the public internet with a password a child could guess, and what they achieved was digital graffiti. The incident is real and worth taking seriously, but the seriousness it warrants is the seriousness of basic hygiene, not the seriousness of national cyber-warfare.

Or consider Colonial Pipeline, whose 2021 shutdown became the reference case for critical-infrastructure cyber-risk in the United States. Strip away the panic-buying and the presidential briefings and ask the narrow technical question: how did the attackers get in? Not through a sophisticated assault on pipeline control systems. They got in through a single virtual private network password, on an account with no multi-factor authentication behind it.[5] One password. No second factor. The entry point to an event that disrupted a pipeline that, as of 2021, was reported to carry roughly 45 percent of the U.S. East Coast's fuel supply was a missing control that a small business is expected to deploy on its email.

Place these two incidents beside the threats the fear machine sells, the nation-state superweapon, the air-gap-leaping payload, the surgical manipulation of a physical process, and the divergence is the entire point of this chapter. The marketed threat is exotic; the realized exposure is a default credential and a missing factor of authentication. This is the gap rendered concrete. The systemic vulnerabilities that follow in this chapter are not, for the most part, cinematic. They are the conditions, accumulated over decades and tolerated for understandable reasons, that turn a guessed password or a stolen login into operational consequence. Diagnosing them honestly, as mundane hygiene problems rather than as the raw material of a doomsday film, is the necessary first step toward spending finite defensive resources where the evidence says the danger actually lives.

Legacy Systems: The Accumulating Surface

One of the most significant systemic vulnerabilities in OT environments is the widespread reliance on legacy systems. Many industrial control systems (ICS) and SCADA (Supervisory Control and Data Acquisition) systems were designed and implemented decades ago, long before cybersecurity was a primary design consideration, or even conceived of as a significant concern. These systems were built with a different set of priorities: reliability, availability, and the continuous execution of their core industrial function. The concept of a network perimeter that needed defending against external threats was often alien to their designers. Consequently, these legacy systems frequently operate on outdated hardware and software, utilizing obsolete operating systems that are no longer supported by their original vendors. This lack of vendor support means that critical security patches are either unavailable or cannot be implemented without risking the operational stability of the entire system.

Here, though, the diagnosis must be careful, because legacy is exactly the territory on which the fear machine plants its CVE flag. The longer these systems remain in operation, the more known vulnerabilities are discovered, documented, and catalogued, and the reflexive conclusion is that each newly disclosed flaw is a fresh dagger at the throat of the grid. The Dragos finding deflates that reflex. Most of those documented flaws sit deep inside networks an attacker would have to already own, are not exploitable for the catastrophic scenario, and have rarely figured as the decisive lever in publicly documented attacks. The genuine hazard of a legacy system is therefore rarely the specific

CVE on the advisory; it is the broader, duller fact that an old, unmonitored, poorly understood device is precisely the kind of thing that ends up reachable from somewhere it should not be, configured with a default credential nobody revisited, sitting on a network nobody segmented. Trying to secure these systems is akin to patching a crumbling, ancient dam; some temporary fixes are possible, but the fundamental structural condition, age, obscurity, and neglect, remains. The point is not that legacy systems are safe. The point is that what makes them dangerous is mundane exposure, not the exotic exploitability the patch-panic narrative emphasizes.

Insecure Protocols by Design

Compounding the issue of legacy systems is the use of insecure, proprietary, or outdated communication protocols. Many OT systems rely on protocols developed in an era when networks were physically isolated and considered inherently secure. Protocols like Modbus, DNP3 (absent the DNP3 Secure Authentication extension), and Ethernet/IP (absent the optional CIP Security profile), while functional for their intended purpose, often lack built-in security features such as authentication, encryption, or integrity checking. Commands and data can be transmitted across networks in plain text, susceptible to eavesdropping, manipulation, or injection of false data. An attacker with even limited network access can potentially intercept operational commands, alter setpoints, or inject malformed data.

Notice the shape of this weakness, because it sharpens the chapter's thesis rather than contradicting it. The protocol exposure is not a bug to be patched; it is the protocol working as specified. Exploiting it does not require a nation-state's advanced capabilities or a zero-day, it requires knowledge of an openly documented protocol and access to the network segment where it runs. This is, once again, the mundane reality. The danger is not that Modbus harbors a secret exploitable flaw; it is that Modbus was never built to distinguish a legitimate operator from an intruder, so the whole game collapses back onto the boring question of who can reach the wire. That is a question of segmentation and access, of hygiene, not of cinematic capability. The attacker who can speak the protocol has already won the only fight that mattered, and that fight was decided by network architecture, not by the sophistication of any exploit. This is consistent with Lee's observation that real OT attacks use legitimate functionality rather than vulnerability exploitation: an insecure-by-design protocol turns ordinary, intended commands into the attack itself.

Flat Networks and the Vanishing Air Gap

A pervasive systemic vulnerability across many OT networks is the lack of proper network segmentation. Historically, OT networks were designed as isolated, air-gapped environments, physically separated from corporate IT networks and the public internet. The increasing demand for remote access, data integration for analytics, and operational efficiency has blurred these lines. In many facilities the separation between IT and OT is porous or nonexistent, and even within the OT environment segmentation is often inadequate. Without robust segmentation, a compromise in one part of the network, or on an internet-connected IT system, can propagate into the critical OT environment. A single infected contractor laptop, a compromised corporate workstation, or an exploited internet-facing device can become a pivot point, allowing an attacker to traverse the network and reach control systems. This flatness lets even low-skilled attackers discover and exploit critical assets once they gain initial access. It is like having a single unlocked door into a building where all the most sensitive rooms are located; once inside, the entire building is vulnerable.

Flatness is, in many respects, the master mundane vulnerability, the one that converts every other unglamorous weakness into operational consequence. A default credential matters only because an attacker can reach the device that bears it. An insecure protocol matters only because an attacker can reach the segment where it speaks. The Colonial intrusion began on an IT credential; what determined how far it could have traveled was the architecture between that credential and anything that mattered. And the vanishing air gap deserves a blunt word, because it is one of the field's most durable myths. The air gap is invoked as a security control, but in practice it is notoriously difficult to maintain and quietly bridged by USB drives, vendor laptops, temporary remote connections, and the relentless business pull toward integration. The Cyber Av3ngers PLCs were not behind any gap at all, they were sitting on the open internet. The honest diagnosis is that segmentation, not exploit sophistication, is the variable that most reliably separates a contained incident from a spreading one, and that getting it wrong is a design-and-discipline failure, thoroughly mundane, eminently fixable, and routinely unfixed.

Patching, Default Credentials, and the Economics of Neglect

The challenge of patching and updating OT systems is real, but it must be framed correctly. Unlike IT systems, where patching is routine, often automated, and frequently performed, patching in OT environments is fraught with complexity and risk. OT systems often operate around the clock with no permissible downtime, and the critical nature of their function means that any instability introduced by a patch can have severe consequences. The hardware and software are often custom-built, proprietary, or deeply integrated, so standard IT patching tools are incompatible. Vendor approval is often required and can be slow. The result is that many OT systems run with known, unpatched vulnerabilities, not because operators are unaware of them, but because the operational risk of patching is judged to outweigh the risk of exploitation.

The conventional telling treats that last sentence as an indictment, a dereliction that leaves the door wide open. The diagnostic reframing of this chapter, grounded again in Lee's hostile-witness testimony, is less alarmed and more honest. If disclosed ICS vulnerabilities have rarely been the decisive lever of an actual ICS attack, then the unpatched flaw is, in the great majority of cases, a theoretical exposure rather than an imminent one, and the operator weighing patch-induced downtime against exploitation risk may be making a more rational trade than the compliance auditor admits. The energy spent treating every missing patch as an emergency is energy not spent on the exposures that the evidence actually implicates.

And those exposures are, overwhelmingly, the prevalence of default credentials and weak authentication. When OT devices are deployed, they are routinely configured with default usernames and passwords that are widely known or trivially discoverable, and in many cases these credentials are never changed, or are changed to something easily guessed. The focus during deployment is on getting the system operational, not on hardening it. An attacker who reaches the device can often simply use the credential the manufacturer shipped. The absence of multi-factor authentication means a single compromised login grants access to critical control functions. This is the vulnerability that requires the least sophistication to exploit, accessible to curious hobbyists and organized actors alike, and it is, not coincidentally, the one the realized incidents keep returning to. The Cyber Av3ngers needed "1111." Colonial fell to a dormant VPN account without multifactor authentication whose reused password enabled the intrusion that led the pipeline operator to proactively shut down operations.[6] If a single lesson runs through the actual record rather than the marketed one, it is this: credential hygiene and basic authentication, the dullest disciplines in the catalog, are where the attacks actually land. A field that spent on default-credential audits and multi-factor rollouts a fraction of what it spends chasing nation-state superweapons would be addressing the exposures the evidence names.

Physical Access as a Mundane Vector

The physical security of OT systems is often an overlooked systemic vulnerability, and it belongs in this chapter precisely because it is so unglamorous. In many industrial settings the physical access controls to areas housing control systems, network infrastructure, or human-machine interfaces (HMIs) are lax. Doors may be left unlocked, badges may not be strictly enforced, and sensitive equipment may sit in areas with limited surveillance. This can be exploited by insiders, by opportunists, or by an attacker who simply walks in. Physical access lets someone plug in a malicious USB drive, tamper with hardware, or connect directly to accessible ports, bypassing digital defenses entirely. The assumption that digital defenses suffice breeds complacency about the physical, creating a critical blind spot.

Consider the tangible assets themselves. Servers, programmable logic controllers (PLCs), human-machine interfaces, and network switches are physical objects that, if compromised in the flesh, can mean a complete loss of control. Unauthorized access to a server room can let an attacker install malicious hardware such as a keylogger or a network tap, disable critical components, or disconnect essential links, none of which any firewall can prevent. A PLC is the brain of many industrial processes; someone with physical access to it can reprogram it, alter its logic, or simply switch it off, bringing a plant or utility to a standstill. And the much-touted air gap, the theoretical isolation of critical OT networks from external ones, is precisely what a physical breach eliminates: it is notoriously hard to maintain even against software, and against a person standing at the rack it offers nothing at all.

The human dimension of physical security is equally mundane and equally decisive. Access control systems are only as good as their implementation. Tailgating, an unauthorized person following an authorized one through a secured door, remains a simple and effective way in. Weak visitor management, thin background checks for personnel with physical access, and insufficient monitoring of sensitive areas all add up, and in older or more remote facilities the emphasis on physical security is often comparatively lax, a holdover from an era when the imagined threats were vandalism and theft rather than cyber-physical sabotage. Third-party contractors deepen the exposure: essential for specialized skills, they often receive privileged physical access, and without rigorous vetting and supervision a contractor can leave a door ajar, briefly disable a camera, or install a small clandestine device in the course of legitimate work. The convergence of IT and OT compounds this too, a compromised corporate workstation can yield the facility layouts, security schedules, or access credentials that make a physical intrusion easy to plan.

The principle of defense in depth, a cornerstone of cybersecurity, applies with full force to the physical realm: multiple layers so that a breach of one leaves others intact. For OT that means robust barriers, sound access control, continuous surveillance, diligent vetting, and a security culture that values physical and digital integrity alike, extended not only to control rooms and data centers but to remote sites, substations, and anywhere critical hardware lives. The argument that sophisticated cyberattacks are the primary threat tends to drive over-investment in digital defense at the expense of these basics, and that is a misallocation. A determined adversary who can walk in physically can bypass years of expensive cyber investment with relative ease, while the cost of strong physical measures, reinforced perimeters, multi-factor controls on physical access, secure enclosures, comprehensive video, is often a fraction of advanced cyber tooling and a more effective deterrent against a whole class of attacks.

The diagnostic point, then, is that the most effective "exploit" of a control system is frequently a walk through an unlocked door, an act requiring no exploit at all. It is the air-gap myth in physical form: the theoretical isolation that a single propped door, a tailgated badge, or an unsupervised contractor erases. This is not the stuff of cyber-war documentaries, which is exactly why it is underdefended relative to its real contribution to risk, and why true resilience, as the closing chapters will argue, requires safeguarding both the bits and the atoms.

The Long Lifecycle and the Unknown Asset

The very nature of OT system deployment and lifecycle management contributes to systemic vulnerability. Many OT systems have an extended lifespan, often operating for fifteen, twenty, thirty years or more. Across that long lifecycle the technology landscape, the threat environment, and security practice all evolve dramatically, while the systems themselves may remain largely unchanged, leaving them perpetually behind the curve. The cost and complexity of replacing or significantly upgrading industrial systems are often prohibitive, so organizations are forced to maintain aging infrastructure. This is not, for the most part, a story of careless operators; it is a story of the economics and engineering realities of the industrial sector, of systems built for longevity and extreme operational demand. (Chapter 11 takes up what governance reform can do about this condition, and Chapter 13 what resilience engineering can; the work of this chapter is only to name it honestly as a structural fact rather than a moral failing.)

Compounding it is the lack of comprehensive asset inventory and visibility. Organizations often have an incomplete or inaccurate understanding of the devices, software, and network configurations in their OT environments. Unknown or unmanaged devices sit on the network running outdated software and reachable via default credentials. Without a clear picture of what is connected and how it is configured, it becomes impossible to implement effective controls, monitor for threats, or respond to incidents. This "unknown unknowns" problem is a direct consequence of asset management deprioritized in favor of uptime and production goals, and it is mundane in the deepest sense: the most dangerous device in many OT environments is not the one with a famous vulnerability but the one nobody knew was there. The Cyber Av3ngers found internet-exposed controllers; somewhere, someone did not know, or did not track, that those controllers were reachable at all. Visibility is the precondition for every other control, and its absence is the quietest, most consequential weakness on this list.

The Human Element, Honestly Stated

Finally, the human element, while often discussed in the context of phishing and social engineering, also represents a systemic vulnerability rooted in the design and operation of OT. The workforce operating and maintaining these systems may not have a strong cybersecurity background; their training and focus are on operational processes, safety, and engineering. Where dedicated cybersecurity personnel exist, their reach within the highly specialized OT domain can be limited. This can lead to misconfigurations and insecure practices, the default credential left in place, the device quietly plugged into a network it should not touch. The "security is everyone's responsibility" mantra often falls short when the primary expertise and responsibility lie elsewhere.

This systemic gap in cybersecurity awareness within the operational culture is a critical vulnerability that can be exploited without any sophisticated attack at all. It is inherent in organizational structure and skill sets, not in the malice of adversaries. And it is the human root of every mundane failure already named: a default password is left because changing it is nobody's job; a segment is left flat because integration is somebody's deadline; an asset goes untracked because uptime is the metric anyone is measured on. These fundamental, systemic weaknesses create fertile ground for exploitation by a wide range of threats, from state-sponsored actors to opportunistic criminals to accidental misconfigurations. The illusion of control stems precisely from focusing on the sophisticated threats while ignoring the foundational cracks, the boring, fixable, perennially deferred cracks, in the system itself.

From Isolated Weakness to Systemic Risk

The preceding discussion has examined the intrinsic vulnerabilities embedded within Operational Technology (OT) systems, weaknesses stemming from legacy architectures, insecure protocols, flat networks, default credentials, and inadequate practice. However, the true amplification of these inherent weaknesses, and the national security implications they carry, is only fully appreciated when we consider the intricate web of interdependencies that defines modern critical infrastructure. The notion of a power grid operating in isolation, or a water treatment facility functioning independently, is a relic of the past. Today these sectors, and all others deemed critical, including telecommunications, transportation, finance, healthcare, and emergency services, are inextricably linked. This interconnectedness, while a testament to the efficiencies of our technologically integrated society, simultaneously creates fertile ground for systemic risks that transcend individual sector vulnerabilities.

The diagnostic value of interdependence, for this chapter's argument, is that it explains how a mundane failure scales into a serious one without requiring any escalation in the sophistication of the cause. A default credential on one operator's device, an insecure protocol on one porous segment, becomes consequential not because the attacker grew more capable but because the system around the weakness is fragile and tightly coupled. The severity of an impact is not solely a function of the attacker's capabilities; it is at least as much a function of the inherent fragility created by interconnection. A single, well-placed, entirely unsophisticated compromise can have far-reaching consequences because the interdependencies were already there to carry it.

Consider the foundational role of the electricity sector. It is the lifeblood of virtually every other critical infrastructure. Without a stable power supply, the modern world grinds to a halt. Water treatment and distribution plants, essential for public health and industrial operations, depend on electricity to power pumps, purification systems, and control mechanisms. A significant disruption to the power grid, whether from a natural disaster, a physical attack, or a cyber incident that compromises control systems, can rapidly cascade into the water sector. This is not hypothetical; it is operational reality. The control systems that manage water flow, pressure, and chemical treatment depend on uninterrupted power to operate their electronic components, sensors, and actuators. When that power is cut, these systems cease to function, potentially leading to a loss of potable water, the inability to fight fires, and public health crises.

The telecommunications sector is equally dependent on robust power infrastructure. Cell towers, data centers, and internet exchange points all require constant electricity to maintain connectivity. A widespread power outage can cripple communication networks, severing the lines vital for emergency response, business continuity, and public information during a crisis. This dependency creates a dangerous feedback loop: a disruption to the power grid simultaneously degrades the ability of telecommunications providers to communicate with their own infrastructure, to coordinate responses, or to provide critical services like emergency calls. And the systems used to manage and monitor the power grid itself rely on telecommunications networks. This mutual reliance means a failure in one can exacerbate and prolong a failure in the other, a domino effect that can rapidly destabilize national operations, regardless of how unglamorous the initiating weakness may have been.

The implications extend into the transportation sector. Modern logistics, air traffic control, and even the functioning of traffic signals depend on electricity and communication networks. A prolonged power outage can bring air travel to a standstill, disrupt the movement of goods and people by rail and road, and produce widespread traffic chaos. The control systems for ports, crucial for international trade, also require reliable power and communication. The economic and social paralysis of major transportation hubs rendered inoperable for an extended period, due to a failure originating in a different, seemingly unrelated sector, ripples across supply chains, affecting everything from the availability of food and medicine to the ability of industries to access raw materials.

The financial sector, the engine of the global economy, is also deeply intertwined with these other critical infrastructures. Its operations depend on secure and reliable communication networks for transactions, on data centers powered by electricity, and on the ability to operate sophisticated computing systems. A significant disruption to the power grid or telecommunications could cripple financial markets, with immediate and potentially catastrophic economic consequences. The speed at which financial transactions occur means even brief interruptions can have amplified effects, and a sustained outage could lead to a loss of confidence and widespread economic instability.

This interconnectedness is not merely an academic observation; it is a fundamental characteristic of how modern societies function. The governance and management structures, however, often lag behind this reality. Critical infrastructure sectors have historically been managed in silos, each developing its own regulatory frameworks, operational standards, and security protocols. Interagency bodies and information-sharing initiatives exist, but the level of coordinated governance required to manage these complex interdependencies is often lacking. A vulnerability that might be contained within a single sector, or an attack that might only cause localized disruption, can quickly escalate into a national crisis when it impacts multiple interdependent systems. The attacker need not mount a sophisticated, multi-sector campaign; a single, well-placed, mundane compromise can have far-reaching consequences because of preexisting interdependencies. (The governance dimension of this problem is the subject of Chapter 11, which treats the same systemic weaknesses through the lens of reform; here the aim is only to establish that fragility, not attacker brilliance, is the multiplier.)

The challenge is compounded because many of these critical systems were designed and built when cyber threats were nonexistent or nascent. The primary focus was operational reliability, resilience against physical threats, and efficiency. Cybersecurity was an afterthought, if it was considered at all. When these systems are now integrated and interconnected, vulnerabilities once isolated within a single sector propagate rapidly. If a power utility implements robust cybersecurity for its operational network but the water utility it supplies has a weaker posture, a compromise in the water utility's network could be used as a vector to impact the power utility even if the power utility's own defenses are strong. The weakest link can bring down the chain, and in a highly interconnected environment that weakest link can be anywhere, which is to say it is usually wherever the most mundane hygiene was neglected.

The adoption of smart technologies and the Industrial Internet of Things (IIoT) in critical infrastructure, while promising greater efficiency and real-time monitoring, also amplifies these interdependencies and introduces new avenues for cascading failure. As more sensors, actuators, and control systems are connected to networks, and as data is increasingly shared across sectors and with external partners, the attack surface expands. A compromised IIoT device in a seemingly low-criticality system can provide a bridge into more sensitive control networks, and the real-time nature of these interconnected systems means an anomaly or a malicious command can propagate almost instantaneously, leaving little time for human intervention. The lack of standardized security frameworks across sectors deepens the problem: some sectors maintain mature programs while others rely on outdated practice or lack the resources and expertise to implement effective controls, so a weakness exploited in a less secure sector can have direct consequences for a more secure one through operational reliance.

The economic realities of maintaining and upgrading critical infrastructure play a significant role here, as they did in the long-lifecycle problem above. The cost of replacing aging systems or implementing comprehensive cybersecurity upgrades across entire sectors can be astronomical, which leads to a prioritization of operational needs over security investment, especially when the perceived threat is abstract. But the interconnected nature of these systems means a localized threat, amplified through interdependence, can become a national crisis with far greater economic and societal costs than the proactive investment would have entailed. The illusion of control persists when each sector believes it is secure within its own domain while failing to recognize that its security is fundamentally dependent on the security of the sectors it relies upon.

In essence, interconnectedness transforms individual, often mundane, vulnerabilities into systemic threats. A weakness in one area, however minor it looks in isolation, can have disproportionately large and widespread consequences. The national security posture of a country is therefore not merely a reflection of the strength of its individual critical infrastructure sectors, but of the resilience and security of the entire interconnected ecosystem. The diagnosis is now complete in one half: the weaknesses are foundational, mostly unglamorous, and dangerous chiefly because they sit inside a fragile, tightly coupled system. The remedies, governance reform and resilience engineering, are the work of later chapters; what this chapter has insisted upon is that we are diagnosing the right disease before we prescribe.

The Supply Chain: A Real Risk, Honestly Sized

The narrative of vulnerability would be incomplete without examining another pervasive and often underestimated vector: the supply chain. The interconnectedness that defines modern OT infrastructure, while enabling efficiency and innovation, simultaneously creates a vast and intricate web of dependencies on external vendors, manufacturers, and service providers. Each link in this globalized chain represents a potential point of entry, a vector through which malware, backdoors, or exploitable weaknesses can be introduced into the heart of critical systems. It is worth pausing on the supply chain here precisely because it is the one place where the mundane diagnosis of this chapter meets something closer to the cinematic, and the discipline of the diagnosis is to size that something honestly rather than to inflate it.

The concept of a supply chain in the OT context is multifaceted and extends far beyond the simple acquisition of physical hardware. It encompasses the entire lifecycle of a system, from the initial design and manufacturing of components, through the integration of software, to ongoing maintenance, updates, and the disposal of old equipment. At each stage, reliance on third parties introduces a degree of risk that is difficult to fully quantify or control. Consider the silicon chips that form the backbone of modern industrial control systems, sensors, and communication devices. These components are often sourced from a handful of global manufacturers, each with its own intricate processes and security protocols. A compromise at this foundational level, perhaps through a malicious firmware component or a deliberately weakened cryptographic element introduced during fabrication, can propagate unseen through countless downstream products.

This is the scenario the fear machine reaches for, and it is the place to apply the chapter's discipline most carefully. The infamous Stuxnet worm is invariably invoked here, and it does serve as a genuine reminder that carefully crafted malware can find its way into highly secured OT environments. But notice how Stuxnet actually traveled. Its principal propagation vector into air-gapped environments is well documented as infected USB drives, and it abused stolen code-signing certificates from Realtek and JMicron to make its drivers appear legitimate, which is to say that even the canonical sophisticated supply-chain attack ultimately relied on thoroughly mundane mechanisms: removable media carried across a boundary, and the implicit trust placed in a routine update. The lesson is not that the exotic supply-chain superweapon is impossible, it plainly is not, but that even its most celebrated example reduced, at the decisive moment, to the same unglamorous failures, trust and removable media, that this chapter has named throughout. Sophisticated supply-chain compromise is real but rare, and where it succeeds it usually still rides in on ordinary hygiene gaps.

Beyond the hardware, the software that governs OT operations is equally susceptible. Industrial control systems rely on complex proprietary software, operating systems, and firmware developed by a range of vendors. When these vendors are themselves attacked, or fail to implement secure development practices, their products can become unwitting carriers of malicious code. A software update intended to patch a flaw or add functionality could instead contain a hidden payload designed to establish persistent access, exfiltrate data, or disrupt operations. The challenge is compounded by the long lifecycles of OT systems, which often remain in service for decades, running software no longer actively supported, or developed before modern security practices were understood. Here the diagnosis loops back on itself: the supply-chain risk of a legacy system is, once again, mostly the mundane risk of obsolescence and unpatchable obscurity rather than the implanted-backdoor thriller.

The complexity of the globalized supply chain means that tracing the origin of a vulnerability or a compromised component can be arduous, if not impossible. A single piece of OT equipment may incorporate components from dozens of suppliers across multiple countries, each with varying regulatory oversight and cybersecurity maturity. A vendor may procure sub-components from another vendor, who in turn relies on a third, and a compromise anywhere in this extended chain can ultimately reach the final product delivered to the critical infrastructure operator. This intricate interdependence makes it incredibly difficult for an end-user organization to gain full visibility into the security posture of every element within its systems, which is the supply-chain expression of the asset-visibility problem already diagnosed: you cannot secure, or even reason about, what you cannot see, and the supply chain is where the unseen multiplies fastest.

The reliance on third-party maintenance and service providers further expands the risk, and does so in the most mundane register of all. Technicians, engineers, and support staff from external companies often require remote or physical access to OT systems to perform diagnostics, repairs, and upgrades. If these providers do not adhere to stringent security protocols, or if their own networks are compromised, they can inadvertently or deliberately become conduits for attackers. The credentials used by third-party personnel, the security of their remote access solutions, and the integrity of their tools are all critical. A compromised service account can grant an attacker the same privileged access as a legitimate employee, allowing relatively easy navigation of the OT network. This is, once again, a credential-and-access problem wearing a supply-chain costume; the adversary who targets a less-secure vendor to reach a more-secure target is exploiting trust and login, not a clever flaw. Economic pressures within global manufacturing compound matters: the drive to cut cost and accelerate time-to-market can lead to outsourcing to regions with weaker quality control or the use of cheaper, less secure components, inadvertently creating weaknesses that adversaries can exploit, or that adversaries can deliberately seed by infiltrating a smaller vendor's development lifecycle.

The implications of a successful supply-chain compromise are genuinely far-reaching, and this is the one dimension on which the diagnosis grants the alarm real weight. A single compromised component or software update could affect hundreds or thousands of identical systems deployed across sectors and geographies, creating the potential for widespread, systemic disruption. Unlike a targeted attack on a single facility, a supply-chain compromise offers a force multiplier, maximum impact from a single well-placed exploit. That potential is why supply-chain risk earns more concern than most items in this chapter. But sizing it honestly means holding two facts together: the consequence ceiling is high, and the realized frequency is low, with even the marquee cases riding in on ordinary hygiene gaps. The appropriate response is therefore not panic but the same unglamorous disciplines scaled outward, rigorous vendor vetting and qualification, contractual demands for secure development and vulnerability disclosure, hardware and software integrity checks on receipt, cryptographic verification of software signatures, and the principle of least privilege rigorously applied to third-party access. These are mundane controls for a risk whose tail is not, and their mundanity is the point: the work of securing the supply chain is mostly the work of refusing to extend implicit trust, the same refusal that answers every other weakness diagnosed in this chapter. Chapters 11 and 13 take up the governance and resilience machinery this would require; the diagnostic claim here is narrower, that the supply chain is a real and high-consequence risk routinely realized through low sophistication, and that it should be defended as such rather than as a screenplay.

The Insider, and the Thinness of the Record

Our analysis would be incomplete if it did not confront a category of threat that, while less discussed in the grand narratives of nation-state cyber warfare, possesses a genuine capacity to disrupt these systems: the insider threat. And it is worth being candid at the outset about a habit of the literature that this chapter intends to break. Discussions of insider threats in OT tend to lean on vivid hypotheticals and gestures toward incidents whose details are said to be classified or unconfirmable, and the effect can read as evasive. The honest framing is simpler. The public record of confirmed, well-documented OT insider incidents is genuinely thin. Some details are legitimately classified; many incidents are handled discreetly for the legal, reputational, and human-resources reasons described below and never reach the public domain at all. We will say plainly where the record is thin rather than dressing thin evidence in dramatic language.

The allure of external adversaries, the state-sponsored hacker, the criminal syndicate, dominates cybersecurity discourse and investment. There is a palpable, almost cinematic drama to the idea of an unseen enemy breaching the digital ramparts, and that drama is, once again, part of what the fear machine sells. Yet within many OT environments, some of the most significant dangers originate from within, embodied in individuals who already possess legitimate access. This is also the point at which the insider threat connects directly to Lee's observation that real OT attacks turn on legitimate functionality and stolen credentials rather than vulnerability exploitation. The insider needs no exploit; the insider has the credential and the access by right. Insiders, whether current employees, former employees, contractors, or business partners, hold an advantage external attackers would spend considerable time and resources to acquire: proximity. They understand the networks, the operational nuances, the physical locations of key equipment, and the established procedures. This intimate understanding, coupled with authorized access, can transform a trusted individual into a vector of serious damage, and it does so through entirely mundane means, a known login, a familiar console, an ordinary command issued with hostile intent.

The spectrum of insider threats is broad. At one end lies the disgruntled employee who, facing termination, demotion, or a perceived wrong, decides to exact revenge, and whose knowledge of the system can be weaponized to inflict maximum disruption: which control parameters to alter, which communication lines to sever, which data to delete. The intent here is unequivocally malicious. Along the spectrum lies the negligent insider, not seeking to cause harm but whose carelessness or failure to follow protocol creates exploitable weakness, weak passwords, a clicked phishing link, lost unencrypted data, unreported suspicious activity. A moment of inattention at a control system interface, an unverified command entered into a SCADA system, or the accidental disclosure of credentials can open a door or directly trigger an operational anomaly. And a particularly concerning subset involves individuals unwittingly exploited by external actors through social engineering, the employee who, believing they are following legitimate instructions from a vendor or a manager, becomes an unwitting agent of an outside attacker. These are not exotic mechanisms. They are the human face of the same mundane failures cataloged throughout this chapter.

The impact of insider action on OT can be profound. Unlike external attacks aimed at data exfiltration or espionage, insider threats often target the core operational functionality of critical infrastructure, to disrupt services, damage physical assets, or cause chaos. The speed at which an insider can act, combined with intimate knowledge, can make containment and recovery far more difficult than with external breaches. The plausible scenarios are easy to sketch and worth stating as scenarios rather than as documented fact: an insider with access to a grid's SCADA system manipulating voltage levels and damaging transformers; a negligent insider uploading a corrupted configuration and causing localized instability; an insider altering chemical treatment dosages in a water supply, or manipulating flow rates and pressure; an insider with access to air traffic or rail signaling systems creating hazardous conditions; an insider in finance manipulating trading algorithms or transaction records. We label these as illustrations of mechanism, not as a catalog of confirmed events, precisely because the confirmed record is thin and we decline to blur the line. Insider threats operate in the blind spots of traditional cybersecurity paradigms. Much of the defensive focus is on perimeter, firewalls, intrusion detection, tools designed to keep external threats out and less effective against someone already inside with legitimate credentials. The problem is compounded because many OT systems, built on older architectures, lack robust logging or granular access controls, so detecting anomalous insider behavior and reconstructing who did what and when is genuinely hard. The human element is inherently difficult to control; technology can be patched and configured, but behavior is complex and unpredictable, and organizations often struggle with the cultural work of instilling a security-first mindset, leaving security protocols seen as burdensome obstacles rather than safeguards.

What can be said responsibly about the historical record is bounded. Reports of former employees or contractors deliberately introducing malware or sabotaging systems after departure have surfaced over the years, and such acts, driven by revenge or grievance, have demonstrated a capacity for operational disruption and costly recovery. There is the often-cited, and now dated, case of a former employee at a major U.S. utility who intentionally shut down backup power generators at a power plant, an act of physical sabotage rather than a cyber intrusion, but one whose intent and abuse of access mirror the insider cyber threat. There are incidents, primarily on corporate networks rather than OT directly, in which a disgruntled employee leaked sensitive data and introduced disruptive malware; the immediate target was corporate, but the principle, an insider using knowledge and access to cause harm, applies to OT, and the growing convergence of IT and OT makes the crossover a legitimate concern. Beyond such examples the public record thins out quickly, and we will not manufacture specificity it does not contain.

Attribution and prosecution of insider threats are difficult. Unlike external attacks that may point to a foreign nation-state or a known criminal group, insider acts are internal and often harder to prove, which contributes to a reluctance to disclose them publicly and further obscures their true prevalence. The legal and HR complexities can be daunting, sometimes leading organizations to address the matter discreetly rather than through formal channels. This is itself part of why the public record is thin, and naming that mechanism honestly is more useful than gesturing at secrets.

Addressing the insider threat requires a multilayered approach that extends beyond traditional cybersecurity tools. It necessitates a focus on human capital, rigorous background checks for sensitive roles, comprehensive and regularly updated security awareness training tailored to OT, and robust access control built on least privilege, so that individuals have access only to what their job functions require. Advanced monitoring and anomaly detection are crucial, observing user behavior, network traffic, and system activity for deviations from baseline, with machine learning able to surface subtle patterns, and real-time alerting to mitigate damage before it escalates. The principle of zero trust, that no user or device is implicitly trusted regardless of location or prior access, is directly relevant: in an OT context it means continuously verifying access, microsegmenting networks to limit lateral movement, and tightly controlling all interactions with critical systems. Notice that these remedies, least privilege, segmentation, monitoring, are not exotic countermeasures to an exotic threat. They are the same mundane disciplines that answer every other weakness in this chapter, applied to the person who already holds the keys.

Ultimately, the illusion of control within OT environments is tested by the persistent possibility of the insider. External adversaries present a formidable challenge, but the danger posed by individuals with legitimate access, whether through malice, negligence, or unwitting cooperation, is a real one, and one whose mechanism, the abuse of legitimate access rather than the deployment of a clever exploit, fits the diagnosis this chapter has built throughout. The digital ramparts may be strong against external siege; the vulnerabilities that lie within, in the systems and in the people who operate them, demand a different, quieter, and more disciplined kind of vigilance.

The Diagnosis, Stated Plainly

We have unmasked the systemic vulnerabilities, and the unmasking yields a finding that is the reverse of the one the fear machine prepares us to expect. The illusion of control is sustained from two directions at once. It is sustained when defenders mistake their dashboards and firewalls for safety while the foundational cracks, legacy obscurity, insecure protocols, flat networks, default credentials, untracked assets, undertrained operators, go unaddressed. And it is sustained, just as powerfully, when the narrative mistakes the rare and cinematic for the probable, overstating the attacker's command of an unfamiliar physical process while ignoring the engineered, physics-anchored safeties, the kind that caused Triton's 2017 assault on a safety instrumented system at a Saudi petrochemical facility (reported to be Petro Rabigh) to trigger a fail-safe plant shutdown rather than a catastrophe, that stand between a compromised controller and a catastrophe.

The evidence is uncomfortable for an industry organized around the superweapon. "Its own leading insider tells us that disclosed ICS vulnerabilities have rarely been the lever of an actual ICS attack" (and consider naming the source explicitly). and that the field may worry too much about them. The realized incidents do not feature surgical cyber-physical wizardry; they feature a PLC on the open internet with the password "1111," and a major regional fuel supply (Colonial Pipeline, May 2021) protected by a single VPN credential with no second factor. The vulnerabilities that matter are mundane. They are problems of hygiene, of architecture, of inventory, and of discipline, and they are, for the most part, unglamorously fixable. That is the diagnosis. It is not a comforting one, because it implicates choices about where attention and money go, but it is, the record suggests, the true one, and a true diagnosis is the only foundation on which the governance reforms of Chapter 11 and the resilience playbook of Chapter 13 can responsibly be built.

Chapter 9

TOWARDS A FEDERATED MODEL: ENHANCING SITUATIONAL AWARENESS

The intricate dance of modern industrial operations, from the humming turbines of power plants to the automated assembly lines of manufacturing facilities, relies on a complex tapestry of interconnected systems. Yet, within this highly optimized and increasingly digitized environment, a fundamental gap persists: a profound deficiency in comprehensive situational awareness. This deficiency is not a minor inconvenience; it is a critical vulnerability that undermines the very resilience and security of our nation's vital operational technology (OT) infrastructure. Situational awareness, in its purest form, is the perception of environmental elements and events with respect to time or space, the comprehension of their meaning, and the projection of their status in the near future. For OT systems, this translates to a real-time, granular understanding of not only the digital state of devices and networks but also the physical environment, the operational status of critical processes, and the evolving threat landscape. It is the ability to see the whole picture, to understand how disparate elements interact, and to anticipate potential disruptions before they escalate into crises.

Earlier chapters established the diagnosis: a fragmented governance structure, a sprawl of overlapping mandates, and a marketplace that profits more from alarm than from clarity. The purpose of this chapter is different. It is not to re-litigate the problem but to specify the architecture of a solution, the one structural reform from which the others follow. If the defining failure of the current arrangement is that no entity reliably sees the whole board, then the defining task of any credible remedy is to construct the board, the shared field of view, on which subsequent coordination and reform can operate. That field of view is what this chapter calls a federated model for situational awareness, and the body that maintains it is what this chapter calls the central coordinating entity. Both are defined here, once and definitively; later chapters that address operational response and governance reform build on the structure laid out in these pages rather than reinventing it.

The current state of affairs across many sectors of U.S. critical infrastructure reveals a stark reality: a fragmented and incomplete understanding of operational status and potential threats. The point here is not to catalog that fragmentation again but to be precise about the specific blindness it produces, because the shape of the blindness dictates the shape of the cure. Unlike the more centralized control often found in information technology (IT) domains, OT systems are frequently managed by a diverse array of stakeholders, each with their own mandates, priorities, and technical capabilities. This diffusion of responsibility creates silos of information, making it exceedingly difficult to establish a unified, coherent view of the overall security posture. The result is a landscape where an organization might possess detailed knowledge of its internal network traffic but remain largely ignorant of potential vulnerabilities in its physical perimeter, or conversely, have robust physical security protocols but lack visibility into the cyberattacks targeting its control systems. The deficiency, in other words, is not primarily a shortage of data. Each organization generates an abundance of it. The deficiency is that the data rarely assembles into a picture larger than any single organization can hold, and threats that move across organizational boundaries are precisely the threats that the boundary-bound organization is least equipped to see.

The challenge is compounded by the diversity of OT environments. Each sector—water, transportation, manufacturing—has unique technologies and regulatory frameworks, making a one-size-fits-all approach to situational awareness inherently flawed. Effective awareness requires not only understanding the specific technologies within an organization but also appreciating how these systems interact with broader industrial ecosystems and how they might be targeted by a wide range of adversaries. Without a coordinated effort to bridge these sectoral divides and establish common frameworks for data sharing and threat intelligence, each entity remains isolated in its own awareness bubble, vulnerable to threats that transcend its immediate purview. This is the crux of the matter, and it is worth stating plainly because it is the hinge on which the entire argument of this chapter turns: the modern adversary does not respect the sectoral and organizational seams that fragment our defenses. A reconnaissance campaign probes the energy sector and the water sector with the same toolkit; a vulnerability in a widely deployed programmable logic controller is a vulnerability wherever that controller is installed, regardless of who owns the facility around it. Our defensive posture is partitioned along lines the attacker often disregards or exploits. The remedy must therefore be a structure that reassembles the picture the attacker has and the defenders lack.

The absence of a clear, authoritative entity to coordinate these efforts means that critical information may not reach those who need it most, or it may be presented in formats that are incompatible with existing systems. The result is a piecemeal approach to security, where each entity operates in relative isolation, unaware of the broader patterns of attack or the systemic vulnerabilities that exist across sectors. Information that would be decisive if correlated remains inert because it is scattered, and even when it is shared it often arrives in a form no recipient can act upon. A federated model attacks both failures at once: it gives the scattered data a place to converge, and it imposes the common language that makes convergence meaningful.

The human element is also a critical, yet often overlooked, factor in the struggle for situational awareness. The highly specialized knowledge required to operate and secure OT systems means that there is often a shortage of skilled personnel. Even when data is collected and analyzed, the ability to interpret it accurately and translate it into actionable insights depends on the expertise of the individuals involved. Furthermore, the disconnect between IT security teams and OT operations teams, a common organizational challenge, can lead to a lack of shared understanding and a failure to appreciate the unique risks and operational realities of each domain. Bridging this cultural and technical divide is essential for fostering a holistic approach to situational awareness that encompasses both the digital and physical aspects of OT security. A federated model does not conjure expertise that does not exist, but it does change the economics of scarce expertise. When a single skilled analyst correlates a pattern across twenty organizations and the result is shared with all of them, the effective reach of that analyst's work is multiplied many times over. Scarcity of talent is precisely the condition under which pooling makes the most difference, and pooling is precisely what a federated structure is built to do.

The current reliance on threat intelligence that is often generalized and lacks specific context further exacerbates the problem. While broad warnings about emerging malware or sophisticated attack techniques are valuable, they are often insufficient to enable effective defense for a specific OT environment. True situational awareness requires intelligence that is tailored to the organization's specific assets, vulnerabilities, and operational context. This means understanding not just that a particular piece of malware exists, but how it might specifically target the organization's PLCs, HMIs, or SCADA servers, and what the potential operational impact would be. This level of granularity is difficult to achieve without robust data collection and sharing mechanisms that extend beyond the confines of individual organizations and across different sectors. Generalized intelligence is abundant but often insufficient on its own; specific intelligence is decisive and difficult to obtain alone. The gap between the two is the gap a federated model is designed to close, by combining the broad pattern visible only at scale with the local context known only on the ground.

The concept of a federated model for situational awareness, which the remainder of this chapter specifies in full, offers a concrete pathway to address these deficiencies. Such a model envisions a collaborative ecosystem where individual organizations maintain control over their own data and systems but voluntarily share relevant threat intelligence and operational status information within a trusted framework. This would allow for the aggregation and analysis of data on a much larger scale, enabling the identification of systemic vulnerabilities, the tracking of emerging threats across multiple sectors, and the development of more effective and coordinated defensive strategies. The transition to such a model, however, requires overcoming significant technical, organizational, and policy hurdles, including establishing common data standards, ensuring robust privacy and security protections for shared information, and fostering a culture of trust and collaboration among diverse stakeholders. Each of these hurdles is addressed in turn below, because a model that cannot survive contact with the practical objections is not a model but a slogan.

The need for enhanced situational awareness in OT environments is not merely a technical challenge; it is a strategic imperative that underpins the resilience and security of national critical infrastructure. The current fragmented and insufficient understanding of threats and vulnerabilities leaves the nation exposed to a wide array of risks, from targeted cyber-physical attacks to cascading failures triggered by seemingly minor disruptions. Addressing this deficiency requires a concerted, collaborative, and forward-thinking approach that acknowledges the complexity of OT systems and the critical importance of a unified, real-time understanding of their operational status and the threats they face. What follows is the blueprint for that approach.

The Federated Model: A Blueprint for Shared Awareness

The federated cybersecurity model is a strategic approach for enhancing situational awareness across diverse operational technology (OT) environments. Its central design choice is to occupy the productive middle ground between two failed extremes. At one extreme lies complete decentralization, the status quo of isolated entities, each defending its own perimeter in ignorance of the wider battlefield; this is the arrangement whose blindness the preceding section described. At the other extreme lies absolute centralization, a top-down command structure that would attempt to manage the cybersecurity of every industrial control system from a single point. The second extreme fails for reasons that are less obvious than the first but no less decisive. A single central authority is unlikely to hold the domain-specific knowledge required to secure a water utility, a chemical plant, and a transmission operator simultaneously; it would become a bottleneck, a single point of failure, and a target of unparalleled value. The federated model rejects both. It balances a unified national perspective with the operational autonomy and unique requirements of individual sectors, and that balance is not a compromise between two bad options but the recognition that situational awareness is itself a two-layered phenomenon, requiring both the macro view that only aggregation provides and the micro context that only local operators possess.

This model is conceptualized around the idea of a central coordinating entity, often envisioned as a federal agency or a designated national authority. It is essential to be precise about what this entity is and, just as importantly, what it is not, because the most common misunderstanding of the federated model is to imagine the center as a regulator or an operator. It is neither. Its role is not to dictate every security measure or to directly manage the cybersecurity of every industrial control system. Instead, its mandate is to establish the foundational elements necessary for effective collaboration and information exchange—the standards, the trusted channels, and the analytic capacity that few if any individual participants could practically build alone. This includes defining overarching cybersecurity standards and best practices that are applicable across critical infrastructure, ensuring a baseline level of security posture that all participants can strive towards. These standards, developed through extensive consultation with industry experts and sector-specific stakeholders, would provide a common language and a shared framework for security operations, making inter-organizational understanding and data correlation feasible. The center, in short, is an enabler and a synthesizer, not a commander.

Furthermore, the central coordinating entity would be responsible for creating and maintaining the infrastructure and protocols that facilitate secure and efficient information sharing. This is the most critical function of the federated approach in the context of situational awareness, and it is worth dwelling on because everything else depends on it. It involves building trusted channels through which entities can voluntarily share relevant threat intelligence, operational status updates, and incident data. The emphasis here is on voluntary participation, recognizing that organizations will only share information if they perceive clear benefits and if their data privacy and intellectual property are adequately protected. This voluntariness is not a weakness to be apologized for; it is the source of the model's durability. A mandate to share, absent trust, produces minimal compliance, defensive over-classification, and data of deliberately low value. A structure that earns participation produces a flow of high-quality intelligence because participants conclude, correctly, that what they put in is exceeded by what they get back. The coordinating entity would act as a trusted intermediary, capable of aggregating, anonymizing where necessary, and correlating this disparate information to generate a comprehensive, macro-level view of the threat landscape that no single entity could achieve on its own. This aggregated intelligence can then be disseminated back to participants in a timely and actionable format, enabling proactive defense and more informed decision-making. The intermediary is, in effect, the membrane through which raw, sensitive local data passes inward and refined, shareable collective insight passes outward.

However, the federated model explicitly delegates significant responsibility and authority to entities at the sectoral or regional level. These entities, which could be industry-specific associations, regional security alliances, or even large conglomerates of interconnected organizations, are empowered to tailor and implement cybersecurity measures that are most appropriate for their unique operational environments. They possess the deep, domain-specific knowledge regarding the particular OT systems, the prevalent threats they face, and the specific operational constraints that govern their day-to-day functioning. This localized expertise is indispensable, and it is the other half of the two-layered awareness the model is built to produce. The central entity might issue a general alert about a new ransomware strain, but the sector-specific entity would be responsible for translating that alert into specific, actionable guidance for its members, considering the unique configurations of their SCADA systems, programmable logic controllers (PLCs), and human-machine interfaces (HMIs). A generic warning becomes a defensive action only after it passes through the interpretive layer that the sectoral entity provides; without that layer, the most sophisticated central analysis arrives as noise.

This delegation of operational control is a cornerstone of the federated model's design. It acknowledges that a top-down, one-size-fits-all approach to OT cybersecurity is inherently flawed. Different sectors have vastly different risk profiles, regulatory landscapes, and technological foundations. A municipality managing its water treatment facilities will have distinct cybersecurity needs and capabilities compared to a large-scale chemical manufacturing plant or a national power grid operator. By empowering sector-specific entities, the federated model ensures that cybersecurity strategies are grounded in practical realities and are responsive to the nuances of each environment. These entities can develop specialized training programs, conduct tailored vulnerability assessments, and implement targeted incident response plans that are highly relevant to their members. The autonomy is not a courtesy extended to participants; it is a functional necessity, because the knowledge required to act well at the edge cannot be held at the center.

The relationship between the central coordinating entity and the sectoral entities is therefore best understood not as a hierarchy but as a dynamic partnership. The central entity provides the overarching framework, the common standards, and the aggregated intelligence, acting as a force multiplier for national security. The sectoral entities, in turn, provide the detailed operational insights, the localized threat context, and the practical implementation of security measures. This symbiotic relationship allows for the leveraging of both national-level strategic understanding and ground-level tactical expertise. It creates a feedback loop where challenges and emerging threats identified at the sectoral level can inform the development of new national standards or alert the central entity to shifts in the adversary landscape, which can then be communicated more broadly. The loop is the mechanism by which the whole becomes greater than the sum of its parts: local observation feeds central synthesis, central synthesis feeds local action, and local action generates the next round of observation, each turn of the cycle sharpening the shared picture.

The effectiveness of a federated model hinges critically on several enabling factors, and these are the practical hurdles promised earlier. First and foremost is the establishment of trust among participants. Given the sensitive nature of operational data and threat intelligence, organizations must be confident that their shared information will be handled securely, confidentially, and used only for the intended purpose of enhancing collective security. Robust data governance policies, clear rules of engagement, and transparent operational procedures are paramount to building and maintaining this trust. Without trust, the voluntary information-sharing component, which is the lifeblood of the federated model's situational awareness capabilities, will falter. Trust, it bears emphasizing, is not a soft or secondary consideration in this architecture; it is load-bearing. Every technical control described later in this chapter—the encryption, the access controls, the audit trails—exists ultimately in service of trust, because the model collapses the moment a participant concludes that contributing data is a liability rather than an asset.

Second, common standards for data formats, communication protocols, and threat intelligence reporting are essential. If each entity collects and reports data in its own unique format, the task of aggregating and correlating this information becomes immensely complex, if not impossible. The central coordinating entity must play a pivotal role in defining and promoting these interoperability standards, ensuring that data from diverse sources can be understood and analyzed collectively. This might involve adopting or adapting existing standards from IT cybersecurity, such as STIX (Structured Threat Information Expression) and TAXII (Trusted Automated Exchange of Intelligence Information), or developing new OT-specific standards where necessary. Standardization is the unglamorous precondition for everything the model promises: correlation at scale is impracticable if the inputs do not share a grammar.

Third, the federated model requires a clear delineation of roles and responsibilities. While the central entity provides guidance and facilitation, and sectoral entities manage their own operations, there must be unambiguous understanding of who is responsible for what at every stage of the cybersecurity lifecycle. This includes responsibilities for threat detection, incident reporting, vulnerability management, and overall risk mitigation. Clarity in these roles prevents confusion, avoids duplication of effort, and ensures that critical tasks are not overlooked. It also helps in establishing accountability, which is crucial for continuous improvement and for building confidence in the model's effectiveness. Because this third factor is the one most often left vague, and because vagueness here is precisely what defeats federated structures in practice, the canonical statement of these roles is set out in its own section below; the reader should treat that section as the authoritative division of labor to which the rest of the chapter refers.

The benefits of adopting a federated cybersecurity model for OT are substantial, particularly in enhancing situational awareness. By aggregating threat intelligence and operational status data from a wide array of sources across different sectors, the central coordinating entity can identify subtle patterns, emerging trends, and systemic vulnerabilities that would be invisible to individual organizations. Consider a coordinated increase in targeted phishing attempts against a particular type of industrial engineer, observed across multiple energy companies. To any one of those companies the activity looks like ordinary background noise, a handful of suspicious emails indistinguishable from the daily churn. Correlated across the federation, the same activity resolves into something entirely different: a sophisticated, multi-sector reconnaissance campaign, the opening move of a deliberate operation. This insight, shared through the federated network, would allow all participating entities to bolster their defenses against this specific threat, rather than each organization discovering it independently, often too late. The pattern that is invisible at the level of the individual is unmistakable at the level of the aggregate, and surfacing exactly that class of pattern is the model's central purpose.

Moreover, the federated model facilitates a more robust understanding of the interconnectedness of critical infrastructure. A cyber-physical attack on a water treatment facility, for instance, could have ripple effects on adjacent industrial operations or even impact the power grid that supplies electricity to the facility. By fostering information sharing across sectors, the federated model enables a more holistic view of these dependencies and potential cascading failures. This allows for the development of more resilient strategies that consider the broader impact of disruptions, moving beyond the confines of individual organizational perimeters. Critical infrastructure is a system of systems, and its failure modes are frequently cross-sectoral; a model that sees only within sectors is poorly equipped to anticipate the failures that matter most.

The federated approach also promotes a more efficient allocation of resources and expertise. Instead of each entity developing its own advanced threat detection capabilities or specialized incident response teams from scratch, the federated model allows for the sharing of these resources and expertise. The central entity might maintain a national-level threat intelligence platform or a specialized cyber-forensics unit that can be leveraged by all participants. Similarly, sector-specific entities can pool resources to develop joint training programs or to fund common cybersecurity tools that benefit all their members. This economy-of-scale approach is particularly valuable for small and medium-sized enterprises (SMEs) within critical sectors, which may lack the resources to implement comprehensive cybersecurity measures on their own. The same logic that makes scarce expertise more valuable when pooled makes scarce capital more effective when pooled, and it is the smallest participants—the rural cooperatives and municipal utilities—who stand to gain the most from a structure that lets them draw on capabilities they could never finance alone.

None of this is to suggest that the implementation of a federated cybersecurity model is without its challenges. The technical complexities of integrating diverse OT systems and legacy infrastructure with modern security monitoring tools remain significant. Ensuring data privacy and intellectual property protection while enabling information sharing requires sophisticated technical controls and stringent legal frameworks. Overcoming organizational inertia and fostering a culture of collaboration among entities that may have historically operated in silos requires persistent effort and strong leadership. The development and adoption of common standards can be a protracted process, involving numerous stakeholders with competing interests. Furthermore, the last-mile problem, ensuring that the aggregated intelligence and guidance from the central or sector entities are effectively translated into practical action at the operational level within each organization, demands continuous reinforcement and support. These are real obstacles, and they are precisely why the architecture that follows is specified in detail rather than waved at in the abstract.

Despite these hurdles, the federated cybersecurity model represents a pragmatic and promising pathway toward achieving significantly improved situational awareness in operational technology environments. It offers a flexible, scalable, and collaborative framework that respects the unique characteristics of different sectors while fostering a unified national defense against increasingly sophisticated cyber and physical threats. By balancing central guidance with decentralized execution, it empowers each entity to play its part effectively, contributing to a collective intelligence that is far greater than the sum of its individual parts, and ultimately bolstering the resilience of the nation's critical infrastructure. This model recognizes that true security in the interconnected OT landscape cannot be achieved in isolation; it demands a coordinated, intelligent, and federated approach.

Having defined the model conceptually, the chapter now turns to its concrete embodiment. The architecture of a federated information sharing platform for operational technology (OT) environments necessitates a confluence of technical, procedural, and governance elements to ensure its efficacy. At its core, such a platform must facilitate the secure and seamless exchange of threat intelligence, vulnerability data, and operational status updates across a heterogeneous ecosystem of participants. This ecosystem typically comprises government agencies responsible for national security and critical infrastructure protection, operators of vital industrial control systems (ICS), and potentially trusted third-party vendors or service providers who have privileged access or expertise. The overarching goal is to cultivate a shared understanding of the evolving threat landscape, a common operating picture, that empowers all stakeholders to make informed decisions, prioritize defensive actions, and enhance the overall resilience of critical infrastructure. This is not about creating a single, monolithic repository of all data, but rather a dynamic, interconnected network that enables targeted and relevant information dissemination. The distinction matters: a monolith would recreate the single point of failure the model was designed to avoid, whereas a network preserves the autonomy that makes participation tenable.

Core Components of a Federated Platform

The components that follow are described at the level of design decision rather than engineering specification. The aim is to make clear what each component must accomplish and why, so that the requirements are intelligible to the policymakers and operators who must commission such a platform, not to prescribe the particular products or protocols that implement them.

A foundational component is a **secure communication infrastructure**. The decision-relevant point is that the entire model depends on participants believing the channel will not betray them, and that belief rests on three properties. The channel must guarantee confidentiality and integrity in transit, which in practice means strong, modern encryption applied end to end. It must verify the identity of every participant before granting access, which means robust authentication, multi-factor at a minimum, paired with granular, role-based controls so that a participant sees only the information it is authorized to see. And it must remain available through disruption, which means redundancy and failover engineered on the assumption that the infrastructure will itself be targeted. These are not exotic requirements; they are the table stakes of any trusted exchange. What makes them consequential here is the asymmetry of failure: a single demonstrated breach of confidentiality would do more to empty the federation than any number of features could do to fill it.

A **standardized data model and ontology** is the second foundational component, and it is the one most often underestimated. It provides the essential common language, defining the structure, format, and meaning of shared information. Without this common language, the vast amounts of data generated by diverse OT environments remain largely unintelligible when aggregated. Standards such as STIX for threat intelligence and TAXII for its transport are vital starting points, but the decision-relevant insight is that IT-derived standards are necessary and insufficient: they must be extended to capture what is distinctive about OT. This includes representing information related to industrial protocols such as Modbus and DNP3, specific hardware components such as PLCs, HMIs, and RTUs, the network architectures found within industrial facilities, and the operational parameters that give a cyber-physical threat its meaning. A well-defined ontology is what allows the platform to link a specific vulnerability found in a particular vendor's PLC to a broader attack campaign targeting that vendor's products across multiple sectors, and to filter signal from noise so that an alert reaches a participant only when it is relevant to that participant's technological footprint. The development and adoption of these standards require significant collaboration among industry stakeholders, cybersecurity experts, and government bodies, and that collaborative labor is itself one of the central entity's defining responsibilities.

The platform requires a robust **data aggregation and correlation engine**. This is the computational heart of the system, responsible for ingesting standardized data, deduplicating it, and applying analytical techniques to identify patterns and trends hidden within siloed observations. The decision-relevant point is that the engine is where the federated model's promise is either realized or lost: it is the machinery that turns scattered observations into the macro-level picture described throughout this chapter. Machine learning and artificial intelligence algorithms can play a significant role here, helping to detect novel threats, surface emerging vulnerabilities, and assess severity, but their most important contribution is correlation across the boundary between the digital and the physical. If a sensor in a power grid begins reporting unusual voltage fluctuations simultaneously with an increase in network traffic from a suspected malicious address, the engine should flag the conjunction as a possible early indicator of a cyber-physical attack rather than treating each event in isolation. The sophistication of this engine directly determines the quality and actionability of everything the platform delivers; a federation with a weak correlation engine is merely an expensive mailing list.

Crucially, the platform must incorporate **tiered and contextualized dissemination mechanisms**. This is the practical expression of the principle, stated earlier, that a common operating picture does not mean every participant sees every piece of data. Information must be filtered, summarized, and routed to the specific needs and responsibilities of each stakeholder. A national security agency may require a high-level overview of all potential threats to critical infrastructure, while a water treatment plant operator needs detailed, actionable intelligence relevant to their specific SCADA systems and vulnerabilities. The decision-relevant requirements are two. First, routing must be rule-driven and role-aware, delivering the right information to the right recipient in the right format. Second, it must be timely, near real-time for fast-moving threats where minutes matter, periodic for strategic assessments where they do not. Delivery channels, secure dashboards, automated alerts, integrations with existing security tools, periodic reports, are implementation details in service of those two requirements. Dissemination is where the value created by aggregation and correlation is finally returned to the participant, and a return that is mistimed or mis-targeted is, from the participant's perspective, no return at all.

A robust **governance framework and policy engine** is also required, and here a boundary must be drawn carefully. The governance discussed in this chapter is the internal governance of the platform itself, the rules that make the data flows trustworthy, not the broader question of regulatory and statutory reform, which a later chapter takes up. Within that scope, the framework encompasses clear rules of engagement, data ownership policies, acceptable use guidelines, and procedures for onboarding and offboarding participants. Defining what information can be shared, under what conditions, and for what purposes is vital for building and maintaining trust. The framework must accommodate legal and regulatory requirements, including applicable privacy and information-protection regimes (such as the Cybersecurity Information Sharing Act of 2015 (CISA 2015; Pub. L. 114-113, Div. N), the Protected Critical Infrastructure Information (PCII) program established under the Critical Infrastructure Information Act of 2002, and, to the extent shared data includes personal information, applicable privacy regimes such as the EU GDPR or the California CCPA) and sector-specific regulations governing sensitive operational data, and the policy engine is what translates these principles into enforceable rules governing data access, usage, and retention inside the platform. The decision-relevant point is that governance is not bureaucratic overhead bolted onto the technical system; it is the codification of the trust on which the technical system depends. Without a clear and well-enforced governance framework, the voluntary participation that is central to the federated model would quickly erode.

The platform must also include **auditing and logging capabilities**. Every interaction with the platform, from data ingestion and analysis to information dissemination and user access, should be logged. The decision-relevant rationale is accountability: comprehensive audit trails make it possible to reconstruct who accessed what data, when, and why, which is indispensable both for forensic investigation after a security incident and for enforcing the governance policies described above. Logging also feeds **continuous monitoring and analytics capabilities**, which extend beyond watching for incoming threats to watching the federation's own health, the volume and velocity of data, the accuracy of predictions, the speed of dissemination, and, most importantly, the measurable effect of shared intelligence on participants' security postures. This last measurement is what allows participants to confirm that the return on their contribution is real, and it is therefore part of the trust machinery as much as the security machinery. A federation that cannot demonstrate its own value to its members will not retain them.

Finally, effective **user interface and user experience design** is critical for adoption, and adoption is not a secondary concern but the condition of the whole enterprise. The decision-relevant point is blunt: even the most sophisticated backend will falter if the people who must use it find it cumbersome, because a tool that is not used produces no shared awareness regardless of its underlying power. Clear visualizations, customizable dashboards that let users prioritize what is relevant to their role, interactive maps of geographically distributed threats, timelines of attack campaigns, and intelligible risk scoring all serve a single end, which is to let a busy operator grasp a situation and act on it without extensive training. Usability, in a voluntary federation, is not a nicety; it is the difference between a platform that becomes indispensable to OT security professionals and one that is logged into once and abandoned.

An Implementation Blueprint

Translating the federated model from concept to reality requires a clear implementation blueprint. The following table outlines the key components, constraints, and metrics necessary for a successful rollout. Each entry answers a question that participants and funders will inevitably ask before they commit, and the table is best read as the minimum set of commitments that must be settled before the first byte of data is ever shared.

Component	Description
Minimum Data Set	Defines the essential information participants must share. This includes Indicators of Compromise (IOCs), vulnerability disclosures for specific OT hardware and software, anonymized network telemetry showing anomalous patterns, and physical security alerts such as unauthorized site access. Defining the floor precisely is what prevents the federation from degrading into a forum where everyone consumes and no one contributes.

Component	Description
Data Standards	Mandates common formats to ensure interoperability. This would leverage existing standards like STIX and TAXII for threat intelligence and ISA/IEC 62443 for industrial automation and control system (IACS) security requirements, with extensions developed for sector-specific protocols and devices. The standard is the grammar without which aggregation is impossible.
Privacy & Liability	Establishes rules to protect sensitive information and encourage participation. This includes a clear legal framework for data anonymization, rules of engagement defining how data can be used, and liability protections for entities that share information in good faith. Good-faith liability protection is the single provision most likely to convert a reluctant participant into a willing one.
Funding Model	A hybrid public-private funding model is most viable. Initial infrastructure development and central coordination could be federally funded, for example through CISA or DOE grants. Ongoing operational costs could be covered by a tiered subscription model for private sector participants, scaled to size or usage, so that the smallest entities are never priced out of the awareness they most need.
Success Metrics	Defines how to measure effectiveness. Key Performance Indicators would include time-to-detection for sector-wide threats, the number of participating entities, the volume and quality of shared intelligence, and a quantifiable reduction in successful attacks on member organizations. These metrics are not merely managerial; they are the evidence base that sustains voluntary participation over time.

The blueprint above settles the terms of engagement. What it deliberately leaves open is the precise allocation of duties among the parties, and that allocation is consequential enough to warrant its own treatment. The effectiveness of any federated structure hinges on a meticulously defined allocation of roles and responsibilities. Without this clarity, the very architecture designed to enhance collaboration and situational awareness can devolve into confusion, inefficiency, and a dangerous paralysis of action. In the context of safeguarding operational technology (OT) within critical infrastructure, this delineation is not merely a matter of organizational neatness; it is a critical requirement for national security and economic stability. A federated model, by its nature, brings together disparate entities, government agencies, private sector operators, and potentially international partners, each with unique mandates, capabilities, and operational environments. To ensure seamless information flow and coordinated response, these entities must understand precisely what is expected of them, what information they are obligated to share, and what actions they are empowered to take. The section that follows is the canonical statement of these roles. It defines, once, the division of labor between the central coordinating body and the constituent entities; the case study that closes the chapter does not restate these roles but shows them operating under pressure.

Roles and Responsibilities

The Central Coordinating Body At the apex of the federated model sits a central coordinating body, the entity responsible for orchestrating the network's operations and ensuring strategic alignment. This entity, which could be an existing government agency with a mandate for critical infrastructure protection or a newly established interagency task force, serves as the linchpin of the entire operation. Its primary responsibility is to establish and maintain the overarching framework within which information sharing and collaborative action can occur. This begins with **establishing and enforcing governance and policy**, the foundational rules that dictate how the shared system operates. The central body must define the rules of engagement for the federated network, developing clear policies on several critical pillars: data ownership, acceptable use of shared information, data classification standards, and the criteria for participation. These policies are not static; they must be regularly reviewed and updated to reflect the evolving threat landscape, technological advancements, and changes in the regulatory environment. The central body acts as the custodian of these policies, ensuring that all participants understand and adhere to them, thereby fostering a predictable and trustworthy environment.

Beyond policy, the central coordinating body is responsible for **facilitating interagency coordination and collaboration**. This involves more than just setting up communication channels; it requires actively fostering relationships and understanding among government agencies, private companies, and research institutions. This entity must bridge the silos that exist between government and industry, and even between different government departments. It acts as a neutral convener, ensuring that all voices are heard and that the needs and concerns of different sectors are taken into account when developing operational tools and information exchange standards. This can manifest in organizing regular working groups, symposia, and training exercises that bring together representatives from various organizations. The goal is to build a shared sense of purpose, the cultural foundation without which the technical foundation is inert.

Furthermore, the central coordinating body bears the responsibility for **defining and maintaining the technical standards and interoperability requirements** of the federated platform. Individual entities may possess their own IT and OT systems, and a common foundation is non-negotiable for interoperability, the ability of disparate systems to connect and exchange data effectively. Achieving this demands stringent requirements for data formats, communication protocols, and security standards that ensure information can be exchanged and understood across the network. The central body's role is not necessarily to build or operate all underlying technology. Rather, it must ensure that any technology adopted or integrated by participants meets the necessary interoperability and security benchmarks, defining the specifications for APIs, data schemas, and security certifications that participating entities must meet. This is the responsibility that gives the ontology and the data standards described earlier their authority; standards without a custodian risk becoming suggestions.

A critical function of the central coordinating body is **aggregate threat analysis and strategic intelligence generation**. Individual entities are best positioned to report on threats within their specific domain. However, the central body possesses a unique vantage point. It synthesizes this information into a broader, strategic picture, collecting, analyzing, and correlating threat intelligence from all participants and identifying overarching trends, emerging attack vectors, and systemic vulnerabilities that transcend individual organizations or sectors. The aggregated intelligence is then disseminated back to the participants in a digestible and actionable format, providing them with the context needed to prioritize their own defensive efforts. This includes generating early warnings, threat assessments, and recommendations for mitigation strategies applicable across critical infrastructure. This function is the reason the central body exists; it is the operational embodiment of the macro-level awareness that no participant can generate alone.

The central coordinating body also plays a role in **supporting incident response coordination**. When a significant incident impacts multiple entities or critical infrastructure sectors, the central body provides the shared situational picture and the trusted communication backbone on which a coordinated response can be mounted, synchronizing the flow of information, surfacing best practices for containment and recovery, and ensuring that lessons learned are captured and fed back into the network. The detailed mechanics of operational response, the joint command arrangements, the inter-organizational agreements, and the conduct of a live, multiparty response, are the subject of a later chapter; the point to register here is narrower and architectural. The coordinating body's contribution to a crisis is the same contribution it makes outside of one, namely the shared awareness and common operating picture that this chapter has been describing, now operating at the tempo a crisis demands.

Finally, the central coordinating body is responsible for **oversight and continuous improvement of the federated model**, the implementation of policies that ensure accountability and transparency. This involves monitoring the performance of the platform, evaluating the effectiveness of information sharing, and gathering feedback from participants to identify areas for enhancement. It conducts regular assessments of the program's effectiveness, identifies any governance gaps or policy deficiencies, and champions initiatives to strengthen the overall resilience of critical infrastructure through collaboration. This commitment to continuous improvement is essential for maintaining the relevance and efficacy of the federated structure in the face of a dynamic threat environment, and it closes the loop with the monitoring and analytics capabilities described among the platform's components.

Constituent Entities The constituent entities are the organizations that make up the network, whether they are government agencies responsible for specific sectors, private sector operators of critical infrastructure, or trusted third-party service providers. Their responsibilities are equally vital and are designed to feed into the overarching goals managed by the central coordinating body. If the central body supplies the macro view, the constituent entities supply the micro view and the action; neither layer is dispensable.

The most fundamental responsibility of each constituent entity is **implementing and maintaining robust cybersecurity measures within its own operational environment**. This is the bedrock of the federated structure. Robust cybersecurity measures include a multilayered defense strategy, proactive threat detection, rigorous vulnerability management, and a commitment to resilience. Each entity is responsible for securing its own OT systems, networks, and data according to established security frameworks and any relevant regulatory requirements, deploying and managing security controls, conducting regular vulnerability assessments and penetration testing, and ensuring that its internal defenses are strong enough to protect against common threats. A weak link within the network can compromise the integrity of the entire federated system, which is why the baseline obligation precedes every sharing obligation: a participant must be reasonably defensible before it can serve as a trustworthy node.

Closely tied to internal security is the responsibility for **timely and accurate reporting of cyber threats, incidents, and vulnerabilities**. Constituent entities must share relevant information with the federated platform, including indicators of compromise (IOCs), details of cyber attacks they have experienced or detected, and information about vulnerabilities they have discovered in specific OT components or systems. The accuracy and timeliness of this reporting are paramount, as inaccurate or delayed information leads to misallocated resources and a distorted understanding of the threat landscape. This reporting must adhere to the standardized data formats and protocols established by the central coordinating body so that the information is immediately usable. Reporting is the input side of the feedback loop; the macro view is only ever as good as the local observations that feed it.

Furthermore, constituent entities are responsible for **providing sector-specific insights and context**. While the central body synthesizes information from a broad perspective, individual entities possess deep knowledge of their particular operational environments, technologies, and unique threat profiles. This specialized knowledge is indispensable for interpreting threat intelligence and understanding its potential impact. An entity operating a nuclear power plant will have a very different understanding of a specific vulnerability than an entity managing a telecommunications network. By contributing their sector-specific expertise, constituent entities help to contextualize the shared intelligence, making it more relevant and actionable for themselves and other participants. This also includes sharing information about the specific OT assets they manage, including their make and model, their software versions, and their network architecture, which is crucial for accurate threat correlation, because the engine can only connect a vulnerability to a victim if it knows where the vulnerable asset is deployed.

Each constituent entity must also take ownership of **implementing and acting upon shared intelligence and mitigation strategies**. Receiving information is only half the battle; acting upon it is what truly enhances resilience. When the federated platform disseminates threat intelligence or recommended mitigation actions, each constituent entity is responsible for evaluating that information in the context of its own operations and implementing appropriate defensive measures. This might involve patching vulnerable systems, reconfiguring critical network devices, enhancing monitoring capabilities, or adjusting established operational procedures. The effectiveness of the federated model is ultimately measured by the collective improvement in security posture resulting from these actions; this is the last mile, and it is the participant, not the center, who must walk it.

Another key responsibility is **participating in exercises and training initiatives**. To ensure that the federated network functions effectively, particularly during crisis situations, constituent entities must actively engage in joint exercises, simulations, and training programs designed by the central coordinating body. These activities help to test communication channels, validate response procedures, and build familiarity among participants, fostering a more cohesive and prepared network. Regular participation reinforces the operational readiness of the entire federated system, so that participants are not exercising core network functions for the first time during an actual crisis.

Constituent entities also have a role in **providing feedback for platform improvement**. The central coordinating body relies on input from its participants to refine the platform's functionalities, policies, and operational procedures. Constituent entities should actively provide feedback on their experience with the platform, highlighting what works well, what needs improvement, and any emerging challenges they face, including any technical issues encountered with data sharing or platform access. This feedback loop is essential for ensuring that the federated model remains relevant, user-friendly, and effective in meeting the evolving needs of its members.

Finally, each constituent entity must ensure its **compliance with data sharing agreements and confidentiality requirements**. The information shared within a federated network is often sensitive. Constituent entities are bound by the agreements they sign upon joining the network, which typically stipulate how shared data can be used, stored, and protected. Maintaining confidentiality and adhering to these agreements is critical for building and sustaining trust among all participants. Any breach of these agreements can have serious repercussions, undermining the entire collaborative effort, which returns the argument to where it began: trust is the substance of which this entire structure is made.

The clear demarcation of these roles and responsibilities is vital for overcoming the inherent challenges in the current U.S. governance structure, which often suffers from fragmentation, from stovepiping, and from a pervasive lack of clear accountability. By establishing a federated model with well-defined duties for both the central coordinating body and the constituent entities, we can create a more cohesive, effective, and resilient framework for protecting critical operational technology. This structured approach ensures that no critical task falls through the cracks while simultaneously preventing burdensome overlaps in effort, leading to a more efficient and impactful cybersecurity posture for the nation's vital infrastructure. This clear division of labor minimizes confusion and maximizes the collective strength of the participating organizations, fostering an environment where shared awareness translates directly into enhanced security. With the roles now fixed, the chapter can show them at work.

Case Study: A Federated Model for the U.S. Energy Sector

The U.S. energy sector, a complex web of public and private assets, presents a compelling case for a federated model, and a useful one for seeing the abstract architecture come alive. Facing a mix of digital, physical, and natural threats, the sector needs a structured framework for information sharing and coordinated action to enhance its ability to anticipate, respond to, and recover from disruptions. What follows is not a fresh recitation of the roles defined above but a walkthrough that places those exact roles into a sequence of concrete events, so that the reader can watch the model perform under load.

Imagine a federated structure for the U.S. energy sector, which we will refer to as the Energy Sector Resilience Network (ESRN). This network would unify the efforts of numerous stakeholders, including investor-owned utilities, public power entities, independent power producers, regional transmission organizations (RTOs), federal agencies such as the Department of Energy (DOE) and the Cybersecurity and Infrastructure Security Agency (CISA), and potentially state-level public utility commissions. The core objective of the ESRN would be to move beyond the current, often fragmented, information-sharing paradigms toward a system that provides real-time, comprehensive situational awareness across the entire energy value chain. In ESRN, the central coordinating body is a dedicated division within the DOE or a newly chartered public-private partnership; the constituent entities are the utilities, producers, and operators that own and run the infrastructure. The energy sector is, helpfully, not starting from nothing. Organizations such as the Electricity Information Sharing and Analysis Center (E-ISAC), operated by the North American Electric Reliability Corporation (NERC), already facilitate some level of threat intelligence sharing, and the DOE's Office of Cybersecurity, Energy Security, and Emergency Response (CESER) already coordinates sector cybersecurity and emergency preparedness. A federated model like the ESRN would not displace these structures but integrate and amplify them under a more cohesive framework, so that information flows seamlessly across organizational boundaries and levels of governance rather than pooling in disconnected reservoirs. With the cast in place, consider how the roles defined in the previous section behave when a real adversary arrives.

The scenario begins with a sophisticated, state-sponsored adversary targeting the U.S. power grid. The first move is quiet. An independent power producer (IPP) in the Southeast notices anomalous activity directed at a specific type of turbine control system, a scattering of probes that, viewed from inside that one company, looks like little more than an elevated background hum. Here the constituent entity's first two responsibilities engage in sequence. Because the IPP has done the bedrock work of maintaining robust internal cybersecurity, its monitoring catches the probes at all rather than letting them pass unremarked. And because it accepts the obligation of timely, accurate reporting, it does not sit on the observation. It promptly transmits the indicators of compromise (IOCs) and the observed attack patterns to the ESRN's central body in the standardized format the central body mandates, the data standards and ontology of the platform doing their unglamorous work so that the report is machine-readable on arrival. Critically, the IPP also exercises its third responsibility, contributing sector-specific insight and context: it attaches detailed telemetry about the affected systems and the precise nature of the attempted intrusion, knowledge that only an operator intimate with that particular turbine control environment could supply. A bare IOC tells the federation that something happened; the IPP's contextual telemetry tells it what kind of something, against what kind of equipment, and how.

Now the central coordinating body performs the function that justifies its existence: **aggregate threat analysis**. Ingesting the IPP's report, the ESRN's correlation engine sets it beside other, individually innocuous signals, faint anomalies that a transmission operator in the Midwest and a distribution utility in the Northeast had each reported as unremarkable upticks in network traffic or minor system irregularities. None of those three organizations, looking only at its own data, could have seen anything alarming. This is the precise blindness the opening of this chapter described, the threat that moves across organizational seams and is therefore invisible to any single boundary-bound defender. But the central body does not look at the data one organization at a time. Correlated across the federation, the three faint patterns resolve into a single sharp one: a coordinated, multi-region campaign against a common class of grid asset. The invisible has become unmistakable, and it has done so only because the macro vantage point exists to perform the synthesis.

What happens next is the return stroke of the feedback loop, the central body discharging its duty to disseminate strategic intelligence back to the edge in actionable form. The ESRN issues a high-priority alert to all participants, and the alert is built for use, not merely for notification. It carries the specific IOCs, names the targeted systems, and lays out a detailed analysis of the adversary's tactics, techniques, and procedures (TTPs). Crucially, it is accompanied by concrete mitigation guidance the platform's tiered dissemination has tailored to the recipients who can act on it: recommended firewall rule updates, the specific software patches to apply, and the steps required to harden the affected OT components against the identified exploit. The existing structures are folded in rather than bypassed. E-ISAC propagates the warning to its broader membership, extending the federation's reach beyond its formal participants, while CESER activates its emergency response protocols to monitor the unfolding situation and stand ready to coordinate federal support. No single utility had to discover the campaign on its own, and none had to invent its own remediation; the federation discovered it once and equipped everyone at once.

At this point the constituent entities discharge the responsibility that ultimately determines whether any of the preceding effort matters: **implementing and acting upon shared intelligence**. Receiving the alert is, as established, only half the battle. Across the federation, participating utilities now evaluate the guidance against their own environments and walk the last mile, applying the patches, pushing the recommended firewall changes, and hardening the named components according to their specific configurations. The RTOs occupy a distinctive position in this phase. As the facilitators of grid operations, they translate the threat intelligence into operational terms no individual utility could, assessing the campaign's potential impact on grid stability and standing up load-balancing measures or emergency generation where the analysis indicates they may be needed. Their capacity to receive real-time intelligence through the ESRN and convert it immediately into grid-protective action is exactly what makes the difference between an isolated intrusion at one producer and a contained, federation-wide defense, and it is precisely the kind of cross-sector load-shedding decision that prevents an attack on a single asset from cascading into a regional failure.

The handling of the live, multiparty response, the secure war rooms in which incident responders from different utilities, government agencies, and cybersecurity firms collaborate in real time, the synchronized containment, the coordination with agencies such as CISA and the FBI, is where the federated platform meets the domain of operational incident coordination addressed in the next chapter. The architectural point to take from the scenario is the one this chapter has argued throughout: the platform's contribution to that response is the shared, correlated, continuously updated situational picture, the common operating picture, on which any coordinated action must rest. Coordination without shared awareness is improvisation; the ESRN's role is to make sure the awareness is already there when the moment arrives.

The same machinery serves threats that have nothing to do with an adversary, which is part of the model's value. Shift the scenario from a state-sponsored intrusion to a major hurricane bearing down on the Gulf Coast. The roles do not change; only the nature of the data flowing through them does. The ESRN now facilitates the rapid sharing of the storm's trajectory, its predicted impact zones, and the real-time operational status of generation and transmission assets in its path. Constituent entities in less affected regions, drawing on the same shared picture, proactively identify resources they can spare, repair crews, spare parts, mobile substations, and stage them for rapid deployment to the hardest-hit areas, the economies-of-scale logic of the federated model expressed as mutual aid. Meanwhile the central body's aggregate situational awareness gives the DOE and other federal agencies a sector-wide view of the disruption's true scope, letting them allocate federal resources where they will do the most good and coordinate coherently with state and local emergency management. A natural disaster partitions damage along geographic lines the way an adversary partitions probing along sectoral lines; in both cases the federated structure reassembles the whole picture that no single participant can hold, and that reassembly is the entire point.

The success of such a federated model would, of course, hinge on addressing the challenges this chapter has not minimized. Building trust among competing private entities and between industry and government would be paramount, the same trust shown above to be the load-bearing element of the entire structure. Clear legal frameworks and contractual agreements would be necessary to govern data sharing, liability, and the use of shared intelligence, the very provisions the implementation blueprint identified as preconditions. Ensuring equitable participation and benefit for all entities, from the largest investor-owned utilities to the smallest rural electric cooperatives, would be essential for broad adoption, which is why the funding model was designed so the smallest participants are never priced out. And investment in secure, interoperable technology platforms would be a significant undertaking, the components specified earlier in this chapter being precisely that investment made concrete.

The potential benefits, however, are commensurate with the effort. A robust federated model for the U.S. energy sector could produce a dramatic improvement in situational awareness, enabling faster detection of threats, more effective response, and a more resilient energy infrastructure capable of withstanding a wider range of disruptive events. It would foster a culture of proactive collaboration, transforming the sector from a collection of independently operating entities into a cohesive and coordinated network capable of collective defense. This would safeguard national security and economic stability and ensure the continued delivery of reliable energy to millions of Americans even under unprecedented strain. The transition to such a federated model represents a significant but necessary evolution in how critical infrastructure is protected, moving from a reactive, fragmented posture to a proactive, integrated system of shared responsibility and enhanced awareness. It is the foundation on which the operational coordination and the governance reforms taken up in the chapters ahead are built, because none of those reforms can function without the shared field of view that a federated model is best positioned to supply.

Chapter 10

IMPROVING INCIDENT COORDINATION AND RESPONSE

The Problem: A Patchwork of Siloed Responses

The diagnosis is, by this point in the book, a familiar one. As Chapter 3 established in detail, U.S. cybersecurity governance for critical infrastructure is a maze of overlapping mandates, sector-specific silos, and crisis-driven institutional accretion, a structure born not of design but of decades of reactive evolution. That chapter traced how responsibility is distributed across a staggering number of federal agencies, layered atop state and local authorities, and further subdivided by sector, and it catalogued the tangible costs of that confusion. There is no need to re-litigate that diagnosis here. The fragmentation is real, it is structural, and it is the water in which every coordinated response must swim.

This chapter takes that fragmentation as a given and asks a narrower, more operational question: when an incident is actually unfolding, across the first chaotic hours and days, how does coordinated response work *mechanically*? Not who should be in charge in the abstract, Chapter 9 has already made the case for a federated model anchored by a central coordinating body, but how the handoffs, the convening, the assignment of authority, and the flow of information actually run on top of that structure once the alarm sounds. The distinction matters. A sound governance architecture and a sound situational-awareness platform are necessary conditions for effective response, but they are not sufficient. Architecture describes who sits where; response mechanics describe what those people do, in what order, with what authority, when the building is on fire. It is entirely possible to have a well-designed organizational chart and a richly instrumented common picture and still fail catastrophically in the first twelve hours of a real event, because no one rehearsed the handoff, no one knew who could authorize a containment action that crossed a sector boundary, and the agreements that should have pre-authorized information sharing were still being negotiated by lawyers while the adversary moved laterally.

The fragmentation Chapter 3 described expresses itself, at the operational layer, as a set of very concrete friction points. When an incident spans multiple sectors, such as a cyberattack affecting the energy grid and financial transactions, the handoffs between distinct coordination mechanisms can be slow and prone to information loss. Each entity may interpret information through its own lens, use different terminology, or operate on different reporting cadences. The energy sector has its own information-sharing mechanisms, such as the Electricity Information Sharing and Analysis Center (E-ISAC), and coordinating entities like the Department of Energy's Office of Cybersecurity, Energy Security, and Emergency Response (CESER). The financial sector has its own critical infrastructure protection programs, and the water sector operates under different mandates. These sector-specific structures are vital for addressing sector-unique challenges, but the seams between them, the precise moment when an incident must pass from one mechanism to another, are exactly where coordinated response succeeds or fails. This chapter is about those seams.

It is worth being precise about why the seams are so dangerous. Agencies like the Cybersecurity and Infrastructure Security Agency (CISA) are designed to facilitate cross-sector coordination, but, as Chapter 3 noted, their authority is often limited to advisory roles and technical assistance. They may lack the direct operational authority to compel action across a diverse set of private entities operating under different commercial imperatives and regulatory oversight. The historical precedent of private ownership and operation of most critical infrastructure fosters a culture in which direct federal command-and-control is viewed with suspicion. The practical consequence, at the response layer, is that coordination must be achieved through *prearranged voluntary structures* rather than through fiat: through agreements signed in peacetime, through joint centers stood up before they are needed, through playbooks drafted and rehearsed when no one is under fire. The whole burden of this chapter is to describe what those structures are and how they operate.

The current incident coordination mechanisms also struggle with the timely and accurate flow of information, and here the operational stakes are sharpest. Reporting requirements can be inconsistent across sectors and even within sectors. Some entities may be hesitant to share information due to concerns about liability, reputational damage, or the potential for adversaries to exploit shared data. Disseminating actionable intelligence to a vast array of stakeholders with varying levels of technical sophistication presents a significant logistical challenge. When an incident occurs, the initial hours are critical. Delays in confirming the nature and scope of the incident, identifying affected parties, and disseminating accurate, actionable information can allow an attack to propagate, cause cascading failures, and amplify the overall impact. The current system, with its multiple layers of communication, translation, and validation, can inadvertently become a bottleneck rather than an accelerator of response efforts. The remedy is not, at this layer, a new diagnosis or a new central body, both already exist on the page. The remedy is a disciplined set of response mechanics that turn the first hours from a scramble into a drill.

Consider where coordinated response most visibly breaks down: the operational response phase of a widespread incident. When a coordinated physical attack strikes multiple substations, or a sophisticated cyberattack disrupts critical communication networks, the response involves a multitude of entities, federal agencies (CISA and other DHS components, FBI, DOE), state and local governments, multiple private sector operators within one or more sectors, and potentially international partners. Coordinating the actions of these diverse actors, each with its own protocols, priorities, and reporting structures, is an immense undertaking. The incident command system (ICS, the emergency-management framework, not to be confused with industrial control systems), widely adopted for emergency management, is typically applied at the tactical or operational level within a single jurisdiction or organization. Its principles, clear chain of command, defined roles, common terminology, unity of effort, are precisely the right principles. The unsolved problem is scaling those principles to a complex, multi-sectoral event across thousands of independent public and private entities. No existing structure, including NIMS and the National Cyber Incident Response Plan, has yet been shown to function reliably as a fully effective 'national ICS' for critical infrastructure resilience integrating public and private operational response at scale, and building one, as a working operational discipline rather than a policy aspiration, is the subject of this chapter.

The Challenge of Cross-Sector Incidents

The inherent complexity of modern critical infrastructure, with its interwoven dependencies, inevitably produces incidents that transcend single-sector boundaries. Managing such cross-sectoral events presents a unique and formidable set of operational challenges, and it is worth dwelling on them concretely, because each one corresponds to a specific response mechanism that the remainder of the chapter will propose. The point here is not to reestablish that fragmentation exists, Chapter 3 did that, but to identify the precise operational failure modes that coordinated response mechanics must be engineered to prevent.

The first operational challenge is the absence of standardized, universally adopted protocols for incident detection, reporting, and response across diverse sectors. Each sector, driven by its specific operational requirements, regulatory mandates, and historical development, has cultivated its own procedures, communication channels, and even terminology. When an incident strikes multiple sectors simultaneously, affecting both energy grids and communication networks, or disrupting transportation systems and financial services, the initial steps of correlating events and establishing a common operational picture become significantly more arduous. At the response layer this is not an abstract interoperability problem; it is the reason that, at hour two of an incident, an analyst in one sector's operations center and an analyst in another's may be looking at the same adversary and not know it, because the indicators are described in different vocabularies and arrive on different schedules.

The core operational difficulty is the seamless aggregation and analysis of information from disparate sources. A manufacturing plant's cybersecurity team might be focused on operational uptime and intellectual property protection, while the energy sector's primary concern would be grid stability and power delivery. The water sector's immediate priority might be ensuring the safety and potability of the water supply. Each entity, operating within its own defined scope, possesses a piece of the puzzle. Without a robust, pre-established mechanism for cross-sector information sharing and correlation, the overarching threat, the interconnected nature of the attack and its potential for catastrophic cascading failures, might not be immediately apparent. The language used to describe the compromise, the metrics for assessing impact, and the urgency with which information is relayed can all differ, creating friction and delaying a unified understanding of the true scope and severity of the incident. The operational answer, developed later in this chapter, is the prearranged information-sharing agreement and the joint analytical cell that fuses these fragments in near real time; the purpose of naming the failure mode now is to make the mechanism legible.

The differing regulatory frameworks governing each sector create a complex web of reporting obligations and authorities that complicate coordinated responses, and during a live incident this web becomes a direct operational drag on the responders themselves. A cyber incident impacting a bank's payment systems can simultaneously trigger reporting obligations to multiple regulators, including Treasury bureaus such as the Financial Crimes Enforcement Network (FinCEN), where the incident involves suspicious financial activity reportable under the Bank Secrecy Act, and prudential regulators such as the Office of the Comptroller of the Currency (OCC), the Federal Reserve, or the FDIC depending on the institution's charter. If that same incident also affects a telecommunications provider that supports the bank's operations, reporting obligations under the Federal Communications Commission (FCC) rules and CISA's purview would come into play. If the attack then disrupts supply chains critical to the financial sector, involving logistics and transportation, a different set of regulations and agencies, such as the Department of Transportation, becomes relevant. This multiplicity means that organizations involved in a cross-sector incident might be dealing with multiple federal agencies, each with its own questions, data requests, and priorities, at the very moment their own responders should be focused on containment. A central operational tenet of this chapter is therefore the single-intake, multi-distribution model: one report, taken once, routed by the coordinating function to every authority that needs it, so that the operator is not simultaneously fighting an adversary and filling out the same incident in seven different forms.

The challenge of synchronized response is the area where cross-sectoral coordination falters most visibly during an event. Even when an incident is recognized as spanning multiple sectors, achieving a coordinated and synchronized response is fraught with difficulty. Each sector, and each organization within a sector, has its own operational priorities, resource constraints, and established response timelines. A widespread cyberattack on electricity substations might require immediate mitigation from utility operators to isolate affected systems, prevent further propagation, and restore power. Simultaneously, that attack might cripple communication networks essential for coordinating the response, affecting emergency services, transportation logistics, and even the ability of financial institutions to process transactions. The synchronization problem is the defining operational problem of cross-sector response: action A taken by one operator, entirely correct from that operator's narrow vantage, can be precisely the wrong action from the standpoint of a neighboring sector, and there is, in the fragmented status quo, no forum in which that conflict is surfaced and arbitrated quickly enough to matter.

The incident command system (ICS), while a valuable framework, is typically designed for hierarchical command and control within a single jurisdiction or organization. Adapting its principles to a complex, multi-sectoral, multi-jurisdictional environment, where public and private entities with different authorities and motivations must collaborate, presents significant scaling challenges. Efforts such as the National Incident Management System (NIMS) and the broader promotion of public-private partnerships have established much of the doctrinal vocabulary, but the practical implementation of a truly unified command structure for cross-sector incidents remains an aspiration rather than a fully realized capability. The gap is not conceptual, the ICS and NIMS concepts are sound, it is operational: the absence of a clear, rehearsed mechanism by which a "national incident commander" function for a given cross-sector event is designated, empowered to direct actions across sector boundaries, and authorized to arbitrate competing priorities in the moment. Closing that operational gap is the heart of this chapter.

The distinct technological environments and operational lifecycles of different sectors pose further challenges for developing common response tools and procedures. The cyber threats and physical vulnerabilities in the energy sector may differ significantly from those in healthcare or transportation. A "one-size-fits-all" approach to incident response is rarely effective. Common operating procedures must therefore be built as a shared core, common terminology, common severity definitions, common escalation triggers, common handoff protocols, wrapped around sector-specific tactical playbooks. The operational design principle is harmonization of the interfaces between sectors, not homogenization of what happens inside each one. The rapid pace of technological evolution within each sector complicates this further, as new systems and applications are constantly deployed, each with its own attack surface and interdependencies, which is one reason the playbooks and agreements described below must be living documents subject to a continuous exercise-and-revision cycle.

The challenge of information validation and attribution in cross-sector incidents adds another operational layer. When an incident originates or spreads across multiple sectors, it can be difficult to quickly and definitively attribute the cause to a specific adversary or to delineate the boundaries of the compromise. Different sectors may have different forensic capabilities, reporting standards for evidence, and legal frameworks for evidence handling. This can delay understanding of the attacker's motives, capabilities, and objectives, all of which are crucial for tailoring an effective response. The ability to rapidly synthesize forensic data from multiple sources, understand the attacker's progression across interconnected systems, and share these findings securely and efficiently across sectoral boundaries is essential but exceedingly difficult to achieve in a fragmented environment. Operationally, this argues for a pre-agreed forensic handoff standard, a common evidentiary format and chain-of-custody expectation, negotiated before an incident so that mid-incident the forensic picture can be assembled rather than reconstructed from incompatible fragments after the fact.

The human element shapes every one of these challenges. The individuals responsible for incident response and coordination within different sectors may have limited exposure to each other's operational realities, regulatory pressures, and technical challenges. Building trust and effective working relationships across these diverse groups requires sustained effort and dedicated resources for joint training exercises, cross-sectoral awareness programs, and informal networking opportunities. Without this human-to-human connection, even the best-designed technical and procedural frameworks can falter under the stress of a real crisis. The reliance on ad-hoc communication during incidents, rather than on deeply ingrained collaborative relationships, increases the risk of misunderstandings, misinterpretations, and missed opportunities for coordinated action. This is why, as we will see, the mutual-aid arrangement and the joint exercise are not soft complements to the "real" technical work of response; they are load-bearing response mechanics in their own right, because in the first hour of an incident people fall back on the relationships and reflexes they already have.

Finally, the concept of "cascading failures" is the defining characteristic of cross-sectoral incidents, and managing it requires a level of foresight and integration that is not always present. A disruption in one sector, such as a prolonged power outage impacting data centers, can trigger failures in other sectors that rely on those data centers, such as financial services, telecommunications, and even emergency response dispatch systems. Understanding these interdependencies and modeling potential failure pathways is a complex analytical task, and during an incident it is also a *coordination* task: the entity managing the originating disruption is rarely the entity best positioned to see the second- and third-order effects in other sectors. Effective cross-sector incident management necessitates a shift from a reactive, sector-by-sector posture to a proactive, holistic response discipline that anticipates cascades and assigns, in advance, the responsibility for watching for them. The mechanics that follow, unified command, prearranged agreements, joint operations centers, response playbooks, mutual aid, and structured handoffs, are precisely the apparatus for converting that aspiration into operational reality.

A Framework for Coordinated Response

If Chapter 9 supplied the architecture, the federated model and the central coordinating body that gives the nation a shared operational picture, then this section supplies the operating manual: the concrete mechanics by which coordinated response *runs on top of* that architecture once an incident begins. The framework rests on a small number of load-bearing components, each of which is an operational instrument rather than a governance principle: prearranged agreements that pre-authorize action and information sharing; a unified command construct adapted from the incident command system; joint operations centers where that command is exercised; response playbooks that turn doctrine into step-by-step action; mutual-aid arrangements that move people and equipment across organizational lines; disciplined handoff protocols for passing an incident between entities mid-event; and an after-action process that feeds every incident back into the next revision of all of the above. Each is treated in turn.

Prearranged agreements: the contracts that make speed possible. The single most important operational lesson of large-scale response is that almost nothing that has to be negotiated during an incident will happen fast enough. The remedy is to negotiate it beforehand, and to encode it in standing agreements. Memoranda of Understanding (MOUs), joint operational plans, and pre-authorized information-sharing arrangements are the connective tissue of coordinated response. Their purpose is to operationalize collaboration: to establish, in writing and in advance, who may share what with whom, who may request what from whom, and what actions are pre-approved so that responders are executing a plan rather than improvising a treaty. For example, an MOU between the Electricity ISAC (E-ISAC) and the Transportation Sector Coordinating Council could preapprove the sharing of specific types of operational data during a grid disruption that impacts transportation networks, thereby bypassing the need to negotiate such agreements under duress. The art of a good response MOU is specificity: it names the categories of data that will flow, the triggering conditions, the named points of contact, the handling and classification rules, and, critically, the liability and confidentiality protections that let a private operator share candidly without fear that today's cooperation becomes tomorrow's lawsuit. Where existing law leaves those protections ambiguous, the framework should drive toward standardized data-sharing agreements that satisfy legal privacy and confidentiality requirements and toward clarified liability protections for entities sharing information during declared emergencies. An agreement that is vague about authority is worse than useless, because it creates the illusion of preparedness while leaving the hard questions to be discovered at the worst possible moment.

Standardized detection, reporting, and a single intake point. Coordinated response begins before command is exercised, with the detection and reporting of the incident itself, and here the framework must mandate a baseline level of commonality. Detection capabilities and reporting procedures currently vary wildly: a sophisticated cyber intrusion might be detected by an Intrusion Detection System (IDS) in one utility, triggering alerts and logs processed according to internal policy, while in another utility a similar intrusion is detected through anomalous operational behavior observed by plant operators, leading to a different set of notifications. This disparity creates delay in the aggregation of threat intelligence and the initial assessment of scope. Harmonization, not homogenization, is the goal: for cyber incidents, encouraging standardized logging formats, common alerting thresholds for critical systems, and interoperable threat intelligence platforms; for physical incidents, standardized sensor data formats and common protocols for reporting physical anomalies with systemic implications. The reporting side is where the operational payoff is largest. Rather than relying on each organization to independently decide which agencies and partners to notify, the framework establishes clear, tiered reporting triggers based on incident severity, impact, and sector, defining what constitutes a "significant incident" requiring immediate cross-sector notification. Reports flow to a single intake point, a clearinghouse function operated by or on behalf of the central coordinating body, that triages, performs initial cross-sector correlation, and alerts stakeholders along predefined escalation paths. The operational principle is report once, distribute many: the operator makes a single notification, and the coordinating function fans it out to every authority with a regulatory or operational equity, sparing the operator the burden of parallel reporting while ensuring no required recipient is missed.

Clear, predefined roles and the question of who is in charge. The mechanics of response collapse the instant there is confusion over who is in charge, who is responsible for what, and who has the authority to decide. The framework must therefore predefine roles across federal agencies, sector coordinating councils (SCCs), Information Sharing and Analysis Centers (ISACs), and individual asset owners and operators. It must name, for each category of incident, the lead federal agency or agencies and clarify the supporting roles of other federal entities (CISA, FBI, relevant sector-specific agencies). It must establish the mandate of SCCs and ISACs in facilitating information sharing and coordination within their respective sectors. And it must outline asset owners' primary responsibilities, detecting, reporting, and mitigating incidents within their own systems, alongside their obligations to share information with designated coordinating bodies. Critically, the assignment of a *lead* is not the assignment of *operational control over private assets*. Within this construct, and absent specific statutory authorities invoked under emergency conditions, the asset owner retains control of its own systems; the lead exercises coordination authority over the collective response. This distinction, coordination authority versus operational control, is the hinge on which public-private response turns, and it must be unambiguous in writing before an incident, not discovered in the middle of one.

Unity of effort and the incident command system, adapted. The governing concept of the response framework is borrowed deliberately from emergency management: the incident command system. ICS supplies the proven grammar of crisis response, a single designated incident commander, a defined command staff, common terminology, manageable spans of control, integrated communications, and a documented incident action plan refreshed on a regular operational cycle. The contribution of this chapter is to adapt that grammar to the cross-sector case. In a single-jurisdiction fire or flood, ICS assumes a unity of command: one commander, one chain. Across thousands of independent public and private entities with their own authorities, strict unity of command is neither achievable nor desirable. The operative principle is therefore *unity of effort*, multiple organizations retaining their own internal command while aligning to a common set of objectives, a shared incident action plan, and a single coordinating authority empowered to sequence and deconflict their actions. This is precisely the gap left open at the end of Chapter 9: the central body provides the architecture and the shared picture; the unified-command construct described here is how a specific incident commander function is designated for a specific event, empowered through the prearranged agreements, and exercised through the joint operations center. ICS also supplies a feature too often missing from cyber response, an explicit, documented operational rhythm: each operational period opens with a briefing, runs against a written action plan with assigned objectives and owners, and closes with a status assessment that feeds the next period's plan. That cadence is what keeps a multi-day, multi-sector response coherent rather than devolving into a swarm of disconnected conference calls.

Joint operations centers and the common operational picture. A command construct needs a place to be exercised. Joint operations centers (JOCs), whether physical facilities or standing virtual coordination hubs, are where representatives from affected sectors and agencies collaborate in real time, analyze evolving information, and jointly develop and approve response strategies. The JOC is the operational embodiment of unity of effort: it is where the incident action plan is built and updated, where cross-sector consultation happens for actions with potential cascading impacts, and where the difficult joint decisions are made, the decision to isolate a compromised segment of a telecommunications network, for instance, which might inadvertently disrupt critical control systems in a neighboring sector, is exactly the kind of decision that must be made *in* the JOC, with the neighboring sector in the room, rather than unilaterally. The JOC runs on the common operational picture (COP): the synthesized, shared understanding of the incident's scope, impact, potential consequences, and ongoing response efforts. The COP is fed by the situational-awareness platform that Chapter 9 described, the federated model is, in effect, the data substrate beneath the JOC's screens, ingesting and correlating technical telemetry, operational status updates, physical sensor data, and threat intelligence feeds, and presenting them through role-based access so that each participant sees what they are entitled to see. The operational point worth stressing is the layering: Chapter 9 builds the picture; this chapter convenes the people who act on it and gives them a room, real or virtual, in which to act together.

Response playbooks: doctrine turned into steps. Roles, agreements, and command constructs are necessary but still abstract; the instrument that makes them executable under stress is the response playbook. A playbook is a pre-written, scenario-specific sequence of actions, decision points, notification triggers, and assigned owners, the operational translation of the framework into "when X is observed, the following people are notified, the following containment options are pre-authorized, the following authority decides, and the following handoff occurs." Good playbooks are tiered to severity, mapped to the reporting triggers, and explicit about the seams between sectors: a cross-sector ransomware playbook, for instance, specifies not only the intra-sector containment steps but the precise moment and manner in which the incident is escalated to the coordinating function and the adjacent sectors are warned. Playbooks must be living documents, drafted in peacetime, validated in exercises, and revised after every real incident and every drill, because a playbook that has never been rehearsed is little more than a hope written down. Their value is highest in the first hour, when adrenaline is high and judgment is degraded, and when the difference between a contained incident and a cascading one is whether the responders are following a rehearsed sequence or inventing one.

Mutual-aid arrangements: moving capability to the need. During a large-scale incident, demand for specialized capability, cyber forensic analysts, operational-technology (OT)-literate responders, spare control hardware, can quickly outstrip any single organization's supply. Mutual-aid arrangements are the prearranged mechanism by which one organization lends qualified personnel and equipment to another under agreed terms, the same logic the fire service has used for over a century and the Emergency Management Assistance Compact (EMAC) has used since its congressional ratification in 1996, applied to critical infrastructure cyber and physical response. Effective mutual aid is more than goodwill; it requires, in advance, agreed definitions of qualified personnel, liability and cost-recovery terms, credentialing so that a borrowed responder can be granted appropriate access quickly, and a dispatching function, naturally exercised through the coordinating body and the JOC, that matches shortfalls to surpluses across organizational and even interstate lines. Without these arrangements pre-negotiated, an operator drowning in an incident must simultaneously beg for help and draft the contract governing that help; with them, the request is a phone call against a standing agreement and the cavalry is credentialed and working within hours rather than days.

Handoff protocols: passing the incident without dropping it. A cross-sector incident is, by its nature, handed off repeatedly, from a local operator to its sector ISAC, from the ISAC to the coordinating function, from one sector's responders to another's as the threat migrates, from the operational responders to the forensic and recovery teams as the phase changes. Every handoff is a moment of maximum risk for information loss, and disciplined handoff protocols are the mechanic that prevents the dropped baton. A good handoff protocol specifies the minimum information set that must transfer, current scope and known indicators, actions already taken, actions in progress, pending decisions and who owns them, and the named receiving point of contact, and it specifies an explicit acknowledgment so that no handoff is presumed complete until the receiver confirms receipt and understanding. This includes the forensic handoff: a pre-agreed evidentiary format and chain-of-custody expectation so that forensic findings assembled in one sector are usable in another rather than incompatible. Handoff discipline is unglamorous, but it is where coordinated response most often fails in practice, and engineering it explicitly, with checklists, named roles, and acknowledgment steps, is among the highest-leverage investments the framework can make.

The after-action process: turning every incident into doctrine. A response framework is not a static document; it must evolve and adapt to new threats, technologies, and organizational changes, and the engine of that evolution is the after-action process. Every real incident and every exercise must culminate in a thorough after-action review (AAR) that documents what happened, what worked, what did not, and why, and that produces concrete, tracked, and implemented recommendations for improvement to protocols, communication channels, roles, agreements, playbooks, and handoff procedures. The AAR is the mechanism by which the seams discovered in one event are closed before the next, by which a playbook that proved brittle is rewritten, by which an MOU that proved ambiguous is sharpened. Exercises feed the same loop and must do more than tabletop discussion: functional and full-scale drills should mimic real operational pressure, testing communication systems, data-sharing mechanisms, incident reporting workflows, and decision-making under simulated stress, and they must deliberately simulate cross-sectoral dependencies, beginning, for example, with a simulated cyberattack on a power grid and tracking its cascading effects through communication networks, water treatment facilities, and financial services, requiring participants from every affected sector to coordinate their responses in real time. The framework must also remain aligned with the legal and regulatory landscape, facilitating alignment with existing regulations where possible and identifying where regulatory clarification or adaptation is needed to enable effective cross-sectoral response. Underlying all of it is trust, fostered not only through well-defined protocols and exercises but through sustained engagement outside of crisis events, regular inter-agency and inter-sectoral meetings, joint training, and initiatives to build personal relationships among the key personnel who will run the response. When individuals know and trust each other, communication flows more freely and collaboration becomes more instinctive during a crisis.

The implementation of such a coordinated response framework requires sustained commitment from both the public and private sectors and significant investment in technology, training, and process development. But the cost of inaction, measured in the potential disruption of essential services, economic losses, and even loss of life, far outweighs the investment required to build a resilient and coordinated response capability. Once established, the framework must be continuously resourced and supported to remain a living, breathing mechanism for safeguarding national critical infrastructure. It is through this structured, collaborative, and adaptive approach, agreements signed in peacetime, command exercised through joint centers, action driven by rehearsed playbooks, capability moved by mutual aid, incidents passed by disciplined handoffs, and every event fed back through after-action review, that the nation moves from a reactive posture to a proactive stance, embedding preparedness in the operational DNA of critical infrastructure resilience and fostering a collective understanding that the security and reliability of one sector are intrinsically linked to the security and reliability of all others.

The Case for a Central Coordinating Authority

The reader who has reached this point may anticipate, given the fragmentation just described, an argument for a central coordinating authority. That argument has already been made. Chapter 9 established the case for a central coordinating body at the apex of the federated model, defined its legislative grounding, and assigned it the architectural functions of governance and policy, interagency facilitation, technical standards and interoperability, and aggregate threat analysis. This chapter does not re-propose that body; it takes its existence as settled and asks the operational question Chapter 9 deliberately left open: when an incident is actually unfolding, how does response *coordination* run on top of that body? The distinction is between the body as a standing institution, Chapter 9's subject, and the body as the convening and command-enabling function during a live event, this chapter's subject. What follows assumes the institution and describes the function.

The first operational function is **convening**. The central body is the entity that, on receipt of a significant cross-sector incident report through the single intake point, decides that the threshold for a coordinated national response has been crossed and convenes the relevant actors, standing up or activating the joint operations center, summoning the named points of contact from the affected sectors and federal agencies, and initiating the first operational period. Convening is a deceptively simple-sounding act that, in the fragmented status quo, frequently does not happen cleanly or quickly, because no single entity has both the vantage point to recognize the cross-sector pattern and the recognized authority to call everyone into the room. Settling that authority on the central body, and rehearsing the convening so that it is a reflex rather than a deliberation, is among the most consequential operational decisions in the entire framework. The speed of convening very often determines the ceiling on the speed of everything that follows.

The second operational function is **designating and enabling the incident command for the event**. The central body does not personally command every incident, nor does it seize the controls of private assets; rather, for a given event it designates the lead, in accordance with the predefined roles, and enables that lead's coordination authority through the prearranged agreements. This is the live-event complement to Chapter 9's standing mandate: the standing body confers, on the designated incident commander function, the practical ability to sequence cross-sector actions and arbitrate competing priorities for the duration of the event. The enabling matters as much as the designating. A lead with responsibility but no recognized authority to deconflict is the precise failure mode that the unmodified incident command system hits at cross-sector scale; the central body's role is to close that gap by lending its legislative legitimacy to the event-specific command.

The third operational function is **information flow management during the incident**. Chapter 9 assigned the central body the standing task of aggregating and correlating threat intelligence; the operational complement is running that flow in real time during an event, ingesting the single-intake report and subsequent updates, maintaining the common operational picture that the JOC works from, fanning notifications out to all authorities with an equity, and pushing correlated intelligence and warnings back to the affected and adjacent sectors fast enough to outpace the adversary's lateral movement. The body acts here as the hub of the report-once-distribute-many model, the node that spares each operator from parallel reporting while ensuring that no required recipient and no endangered neighbor is left uninformed. The operational discipline is tempo: information that arrives after the decision it should have informed is operationally worthless, so the body's information function is judged not by completeness alone but by whether the right picture reaches the right responder inside the window in which it can still change the outcome.

The fourth operational function is **resource and mutual-aid orchestration**. During a large-scale incident, demand for specialized cybersecurity expertise, operational-technology (OT)-literate responders, and essential equipment can quickly outstrip supply. The central body, with the national vantage point Chapter 9 gave it, is uniquely positioned to identify shortfalls and imbalances and to act as the national dispatcher for critical response assets, matching needs to availability across federal reserves, interstate mutual-aid agreements, and private-sector partnerships, so that the right people and equipment reach the right place at the right time rather than being inadvertently siloed within individual response efforts. This dispatching function is the operational expression of the mutual-aid arrangements described earlier: the standing agreements define the terms, and the body, during the event, executes the matching.

The fifth operational function is **post-incident integration**. When the operational phase ends, the central body leads the after-action process across the participating entities, capturing lessons learned and feeding them back into policy, doctrine, playbooks, agreements, and operational plans so that the response to the next incident begins from a higher floor. This is the body's role as the custodian of the continuous-improvement loop at the *response* layer, the operational mirror of the standing governance custodianship Chapter 9 described, and it is what converts a one-off response into durable national capability.

A note on placement is warranted, though it is, properly, a question of the standing institution and therefore largely Chapter 9's. Whether the body is a standalone agency, a division within an existing department, or a council of agency leaders bears on its operational effectiveness during an incident: a standalone agency may offer focus but face integration friction; a divisional home within a department such as DHS may leverage existing interagency relationships but inherit departmental politics; a council may foster buy-in but struggle with the decisiveness that live incident coordination demands. From the operational standpoint of this chapter, the decisive criterion is whichever arrangement best supports rapid convening, clear designation of event command, and decisive arbitration under time pressure. Whatever its form, the body's operational legitimacy with the private operators who own most critical infrastructure is paramount: its convening must be answered, its coordination authority respected, and its information handling trusted, and that legitimacy is built in peacetime, through the agreements, exercises, and relationships described throughout this chapter, not conjured in the middle of a crisis. Legislative mandates should support and protect private-sector participation in national response efforts, through liability protections and incentives for collaboration, so that the operator who answers the body's call in good faith is shielded rather than exposed.

The development and operationalization of this coordinating function would be a significant undertaking, but the consequences of failing to make the standing body operationally real, of having the institution on paper and the convening, command-enabling, information, dispatch, and after-action functions unrehearsed, are far more severe. In an increasingly interconnected and threat-laden world, the ability to mount a unified, synchronized, and resourced national response to critical infrastructure incidents is not a matter of choice but a fundamental requirement for national security and societal well-being. It is the operational apex of the architecture Chapter 9 built: the moment when the shared picture, the standing relationships, and the prearranged agreements are translated, under fire, into a deliberate, coordinated defense of the nation's essential services. The body is the conductor; the framework in this chapter is how the orchestra actually plays together when the performance cannot be stopped.

An Illustrative Scenario: Response Mechanics in the First Seventy-Two Hours

To make the response mechanics concrete, consider the following illustrative scenario. It is hypothetical, a thought experiment designed to show *how coordinated response runs*, who convenes, who holds authority, and how information flows, across the critical first seventy-two hours, not to reestablish that fragmentation exists or to re-argue for a central body. Assume, for the purposes of the illustration, that the framework described in this chapter and the central coordinating body established in Chapter 9 are already in place and operating. The point of the scenario is to watch the machinery work.

Hour zero to hour two: detection and single-point intake. A novel ransomware strain emerges that targets the supervisory control and data acquisition (SCADA) systems commonly used in water treatment facilities and electricity distribution networks. The initial infection occurs within a midsized municipal water authority operating under local governance with limited cybersecurity resources. Its operators detect anomalous behavior in the SCADA environment and, rather than deliberating over which of a dozen agencies to call, they execute the cross-sector ransomware playbook: they begin pre-authorized intra-sector containment and they file a single report to the intake clearinghouse function operated on behalf of the central coordinating body. Almost simultaneously, a regional electricity distributor, more sophisticated in its defenses, detects a closely similar intrusion and files its own single report through the same intake point. Because both reports land at one place in a common format, the correlation that, in a fragmented world, might never have happened, or might have happened only days later after the damage was done, happens almost immediately: the clearinghouse recognizes a shared indicator pattern across two sectors. This is the first and most important difference the mechanics make, the seam between water and energy is bridged at the intake point in hour one, not rediscovered after the fact.

Hour two to hour six: convening and the standup of unified command. Recognizing a significant cross-sector incident, the central coordinating body convenes. It activates the joint operations center, summons the named points of contact from the water and energy sectors, from CISA, the FBI, and the Department of Energy, and, under the predefined roles, designates the lead for the event and enables that lead's coordination authority through the standing agreements. Note precisely what this designation does and does not do: the water authority and the electricity distributor each retain operational control of their own systems, the incident commander function is not seizing anyone's controls, but there is now a single recognized authority empowered to sequence cross-sector actions, to arbitrate competing priorities, and to direct the shared response objectives. The first operational period opens with a briefing against a written incident action plan. The questions that, in the fragmented status quo, would be paralyzing, who is in charge, who decides, who may share what with whom, are not being debated at hour four; they were answered in the MOUs and joint plans signed in peacetime, and the responders are executing rather than negotiating.

Hour six to hour twenty-four: information flow, joint decisions, and mutual aid. The common operational picture, fed by the federated situational-awareness platform of Chapter 9, now shows the JOC a single fused view of both sectors' telemetry, containment status, and the evolving indicator set. Through the report-once-distribute-many model, the initial reports have already been fanned out to every authority with a regulatory or operational equity, so neither operator is simultaneously fighting the ransomware and filing the same incident in parallel forms. Correlated intelligence and concrete mitigation guidance flow back out to both sectors, and, crucially, to adjacent sectors that share the vulnerable SCADA technology but are not yet hit, so that warning outpaces the adversary's lateral movement. The hard cross-sector decisions are made *in the JOC, with the affected parties in the room*: when the energy distributor proposes isolating a network segment to halt propagation, the potential downstream effect on systems that other sectors depend upon is surfaced and deconflicted before the action is taken, not discovered afterward as a cascade. Meanwhile the municipal water authority, short on specialized OT forensic talent, invokes a mutual-aid arrangement; the central body, acting as national dispatcher, matches the shortfall to available analysts from a partner organization, who are credentialed under the standing agreement and working the incident within hours rather than days.

Hour twenty-four to hour seventy-two: handoffs and the transition to recovery. As the incident matures, it is handed off repeatedly, and the handoff discipline is what keeps the baton from being dropped. As the threat is observed probing a third sector, responsibility for that flank is handed to that sector's responders with the full minimum information set, current scope, known indicators, actions taken, actions in progress, pending decisions and their owners, and a named receiving point of contact, and the handoff is not presumed complete until the receiver acknowledges it. Forensic findings assembled by the borrowed analysts at the water authority transfer to the energy sector's investigators in the pre-agreed evidentiary format, so the attacker's progression across the two environments can be assembled into one picture rather than two incompatible fragments. As containment holds, the incident transitions from the operational responders to the recovery and forensic teams through the same disciplined handoff, and the JOC begins to wind down its operational periods. Throughout, the unity of effort holds: dozens of independent public and private entities have retained their own internal command while aligning to a single incident action plan under a single coordinating authority.

After the seventy-two hours: the after-action loop. When the operational phase closes, the central body leads an after-action review across every participating entity. The seams that strained, perhaps a playbook step that proved ambiguous, perhaps a handoff that an acknowledgment step would have caught, are documented, and concrete recommendations are tracked and fed back into the playbooks, agreements, and handoff protocols so that the next cross-sector ransomware event begins from a higher floor. This is the mechanic that turns a single coordinated response into durable national capability.

The contrast with a fragmented response is the entire point of the illustration, and it is a contrast of *mechanics*, not of diagnosis. Nothing in the scenario depends on inventing a new central body or rediscovering that the landscape is siloed; both were settled in earlier chapters. What the scenario shows is that, given the architecture of Chapter 9 and the operating manual of this one, the cross-sector connection is made at intake in hour one rather than missed for days, a recognized authority is convened and empowered in hours rather than argued over, information flows on a single rail rather than through a dozen incompatible reporting channels, scarce expertise is dispatched to the need rather than hoarded, the cascade-inducing decision is deconflicted before it is taken rather than regretted after, and the incident is passed from hand to hand without being dropped. That is what coordinated response *mechanically* looks like, and it is the difference between an incident that is contained in the first seventy-two hours and one that becomes a cascading national event.

From Theory to Practice: The Role of Training and Exercises

The most meticulously crafted plans are ultimately tested not in conference rooms but in the crucible of a real-world incident. To bridge that gap, to transform theoretical preparedness into tangible response capability, a regimen of rigorous, realistic training and exercise is not merely beneficial; it is indispensable. This is where the response mechanics of this chapter, the convening, the unified command, the playbooks, the handoffs, the mutual aid, move from policy documents to ingrained operational competencies. Without consistent and deliberate practice, even the most elegant coordination mechanisms risk becoming brittle, prone to failure when confronted with the unpredictable pressures of a crisis. An MOU that has never been invoked in a drill, a playbook that has never been run under simulated stress, a handoff protocol that exists only on paper, each is a latent failure waiting for the worst possible moment to reveal itself.

The objective of a comprehensive training program is to cultivate a shared understanding and collective muscle memory among all entities involved in incident response. This encompasses a broad spectrum of stakeholders extending far beyond the immediate federal agencies typically associated with national security. It must, by necessity, include the private-sector asset owners and operators who manage the vast majority of the nation's critical infrastructure; their intimate knowledge of their systems, their operational constraints, and their inherent capabilities is vital, and it is their controls that will remain in their own hands during an incident even as they align to a common plan. The program must also integrate the diverse array of emergency responders at the state, local, tribal, and territorial levels, whose on-the-ground actions are critical for immediate containment, mitigation, and recovery, not only traditional first responders like fire departments and law enforcement, but specialized technical teams, public health officials, and emergency management agencies. The aim of exercising this full community together is precisely to rehearse the handoffs between them, because the handoff is where coordinated response most often fails.

The design and execution of these training and exercise regimes must be guided by a principle of progressive realism. This means starting with foundational training that instills common terminology, clarifies roles and responsibilities as defined by the response framework, the distinction between coordination authority and operational control chief among them, and introduces the communication tools and protocols that will be employed during an incident. This foundational layer ensures that, regardless of organizational affiliation, every participant understands the basic operating language and the expected interactions within the coordinated response. The exercises themselves should then escalate in complexity, scope, and intensity. They can begin with tabletop exercises, discussion-based scenarios that walk participants through an incident and let them explore their roles, their decision points, and the operation of the framework, invaluable for identifying gaps in policy, playbook, or understanding at relatively low cost and without the logistics of a full-scale simulation. A tabletop is the cheapest place to discover that a playbook's escalation trigger is ambiguous or that two MOUs contradict each other.

As familiarity and confidence grow, the program should advance to functional exercises that test specific capabilities within the response system. A functional exercise might focus on the activation of the central body's convening and information-fusion function, testing its ability to take a single-intake report, correlate it across sectors, and push warnings outward fast enough to matter. Alternatively, it could test the mutual-aid dispatch process, examining how the coordinating function identifies a shortfall, matches it to a credentialed source, and gets a borrowed responder access under the standing agreement. These exercises are more hands-on than tabletops, often involving live communication systems and the development of simulated operational products, but they typically stop short of physically deploying personnel or equipment.

The pinnacle of this progression is the full-scale exercise, the most demanding, realistic, and comprehensive simulation, designed to replicate the conditions of a major incident as closely as possible. It involves the actual deployment of personnel, the activation of joint operations centers, the use of live communication channels in a controlled environment, and the integration of real-world response actions. A full-scale exercise for a critical infrastructure incident might involve a simulated cyberattack that disrupts a national power grid, requiring the coordinated response of the energy sector, telecommunications providers, financial institutions, federal agencies such as the Department of Energy and CISA, and state and local emergency management. Such an exercise tests the entire chain of mechanics end to end: the single-intake report and cross-sector correlation, the convening and standup of unified command, the operation of the JOC and its common operational picture, the execution of the playbooks, the joint deconfliction of cascade-inducing decisions, the mutual-aid dispatch, the handoffs as the incident migrates and as it transitions to recovery, and the public communication strategy.

Crucially, these exercises must be designed and facilitated with a clear objective: to test the established framework and its mechanics and to identify areas for improvement. The scenarios should be carefully crafted to stress the seams between different organizations and levels of government, exactly the seams this chapter has argued are where coordinated response succeeds or fails. They should deliberately introduce complexities, conflicting information, resource constraints, political pressures, and the cascading effects of failures in one sector on another, forcing participants to invoke the very hand-off, mutual-aid, and joint-decision mechanics the framework provides. The goal is not to achieve a perfect outcome during the exercise but to uncover the vulnerabilities, the points of friction, and the communication breakdowns that might occur in a real event. This requires a robust after-action reporting process: following each exercise, a thorough debriefing and analysis must document what happened, what worked, what did not, and why, leading to concrete recommendations that are tracked and implemented and fed back into the playbooks, agreements, and protocols, the same after-action loop that closes around real incidents.

Building trust and familiarity among participants is another profound benefit of regular training and exercise. In the heat of a crisis, individuals and organizations are more likely to collaborate effectively if they have prior experience working together. They will understand each other's operational cultures, communication styles, and inherent capabilities and limitations. This familiarity reduces uncertainty and fosters a sense of shared purpose. When a cybersecurity professional from a critical infrastructure operator knows the FBI agent they are communicating with has a baseline understanding of industrial control systems, or when a state emergency manager has a pre-established rapport with their counterparts at CISA, the response will be significantly more fluid and efficient. Exercises provide a safe environment to build these relationships and to establish the personal connections that can be invaluable when the pressure is on. This extends to understanding the legal and policy authorities that govern each entity's actions, so that during an incident there is no confusion about who can authorize what, the operational distinction that the framework draws and that the exercises must drill until it is reflexive.

Private-sector involvement in these exercises is a prerequisite for meaningful national preparedness. Critical infrastructure is overwhelmingly owned and operated by private companies. Federal and state agencies can develop plans and establish coordination mechanisms, but their effectiveness is directly tied to the willingness and ability of these private entities to participate. Forcing participation through regulatory mandates, while sometimes necessary, is less effective than fostering genuine partnership. Collaborative exercise design, in which private-sector stakeholders have a voice in developing realistic scenarios and identifying their training needs, can significantly enhance buy-in and commitment, and it reinforces the operational legitimacy the coordinating body depends on. The exercises must be designed to respect proprietary information and operational security concerns while still achieving the realism needed to test the integrated response, which may involve developing simulation environments that mimic operational realities without exposing sensitive data.

The frequency and scope of these exercises must be commensurate with the evolving threat landscape and the dynamic nature of critical infrastructure. A static, infrequent exercise schedule will quickly become obsolete. Threat actors are constantly innovating, and the infrastructure itself is continually evolving through technological upgrades, new deployments, and changes in operational paradigms. Training and exercise programs must therefore be dynamic, adapting to new threats, emerging technologies, and changes in vulnerabilities, through a continuous cycle of scenario development, exercise execution, and after-action analysis and improvement. This is the same living-document discipline the framework demands of its playbooks and agreements, and it requires a sustained commitment of resources, both financial and human, so that the programs can be executed effectively and the lessons learned translated into actual improvements.

The central coordinating body, as established in Chapter 9 and as exercised operationally throughout this chapter, would play a pivotal role in championing and, where appropriate, mandating these training and exercise regimes. It would establish the overall framework, set standards for exercise design and evaluation, and ensure a sufficient tempo of activity across all relevant sectors and governmental levels. It would facilitate cross-sectoral exercises, breaking down the traditional silos that prevent effective collaboration, and serve as the repository for best practices and lessons learned, disseminating that knowledge to all participants and driving continuous improvement in the national response capability. The legislative mandate for such a body should explicitly include responsibility for developing and overseeing a national training and exercise program for critical infrastructure incident response, the training counterpart to its live-incident coordination role.

One particularly challenging aspect of training and exercises is the simulation of cascading failures and complex interdependencies. Modern critical infrastructure systems are highly interconnected; a failure in the energy sector can have profound impacts on telecommunications, water treatment, financial services, and transportation. Exercises that focus solely on individual sectors, or that fail to adequately model these interdependencies, will not prepare the nation for the reality of a major incident, and, just as importantly, they will not exercise the cross-sector handoff and joint-deconfliction mechanics that exist precisely to manage cascades. Scenarios must therefore be designed to explore these interconnected vulnerabilities, forcing participants to grapple with the downstream consequences of disruptions in other sectors. This requires a sophisticated understanding of these interdependencies, often involving advanced simulation tools and data models that can accurately represent the flow of services and the impact of failures across different infrastructure domains.

The human element of response, the stress, the fatigue, the emotional toll of a crisis, cannot be fully replicated in an exercise, but its impact can be better understood and mitigated through deliberate training. Exercises should include elements that test the resilience of response teams, the effectiveness of leadership under pressure, and the protocols for managing personnel welfare. Training on incident command system principles, decision-making under uncertainty, and effective communication in high-stress environments are critical components, precisely because the unified-command construct and the handoff discipline this chapter describes only work if the people executing them can do so while exhausted and under fire. The development of mental health and well-being support mechanisms for responders, often overlooked, should also be part of the training and preparedness discourse, as sustained, effective response requires healthy and resilient individuals.

The role of technology in training and exercises is also rapidly expanding. Advanced simulation technologies, virtual reality (VR) and augmented reality (AR) training environments, and sophisticated data analytics platforms can create highly immersive and realistic training experiences. VR can allow operators to practice responding to equipment failures in a virtual replica of their facility, while AR can provide real-time data overlays and guidance to field personnel during a simulated incident. These technologies can offer cost-effective and safe ways to expose personnel to a wide range of scenarios and to practice complex procedures that would be difficult or impossible to replicate in the physical world. They can also be used to rehearse the operation of the joint operations center and its common operational picture, letting participants drill the information-flow mechanics repeatedly at low cost. But these technological tools must be integrated thoughtfully into a broader training and exercise strategy, ensuring that they complement, rather than replace, the human interaction and collaborative problem-solving that are at the heart of effective incident coordination.

Ultimately, investment in robust training and exercises is an investment in national resilience. This continuous cycle of practice, evaluation, and refinement builds the confidence and cohesion required to protect critical infrastructure. It is not an optional add-on to a national security strategy but a core requirement, and it is the activity that converts the response mechanics of this chapter, and the architecture of the chapter before it, from words on a page into a capability that holds under fire. Without it, the most sophisticated policies remain theoretical, vulnerable to the realities of a crisis.

Chapter 11

REDUCING SYSTEMIC VULNERABILITIES THROUGH GOVERNANCE REFORM

The inherent susceptibility of operational technology (OT) systems is, by now in this book, a settled matter. Chapter 8 set out the diagnosis in full: the vulnerabilities that menace critical infrastructure are not, for the most part, born of malicious genius but are embedded in the systems themselves, a legacy of design choices, operational necessity, and the long, slow drift of technology past the moment of its own obsolescence. That chapter named the conditions honestly and, at several points, deferred their cure to this one. The work of this chapter is to take up that deferral. If Chapter 8 asked what is wrong with the machines, this chapter asks what is wrong with the rules governing the machines, and what it would take to fix them. The distinction matters, because the two failures are routinely conflated, and the conflation produces a particular kind of paralysis: a tendency to treat every systemic weakness as a technical problem awaiting a technical patch, when many of the most consequential weaknesses are, at root, problems of policy structure, of misaligned incentives, and of mandates that overlap, contradict, or simply fail to reach the entities they are meant to protect.

It is worth stating plainly what this chapter is and is not. It is not a fresh inventory of OT's technical frailties; that inventory belongs to Chapter 8 and need not be recited again. It is not a roster of the federal agencies that touch cybersecurity, nor a fresh argument for or against a central authority; the crowded agency landscape and its overlapping mandates are the subject of Chapter 3, and the case for a federated model with a central coordinating entity is built in Chapter 9. What this chapter owns, and what the preceding chapters have been careful to leave to it, is the reform agenda itself: the concrete work of restructuring policy so that it actually reduces the systemic vulnerabilities the rest of the book has documented. That work has several distinct strands, streamlining and harmonizing the tangle of overlapping mandates, reforming how government buys technology so that its purchasing power pulls the market toward security, harmonizing the standards that fracture the OT landscape, regulating for security as a property of design rather than a feature bolted on afterward, and aligning the incentives that today reward speed and cost over resilience. Each is a lever. None of them requires inventing a new threat; all of them require the harder thing, which is changing how the institutions behave.

From Diagnosis to Remedy: What Reform Must Address

As Chapter 8 detailed at length, the systemic vulnerabilities of OT cluster around a handful of mundane, structural realities: aging infrastructure running technology long past its intended lifespan and incapable of supporting modern encryption or patching; systems that are insecure by design, shipping with default or hardcoded credentials, unauthenticated legacy protocols such as classic Modbus, and flat networks that trust all internal traffic; a near-total absence of standardization across a patchwork of proprietary vendors; and a deep interdependency among sectors that turns a single mundane compromise into a cascading, cross-sector failure. That is the diagnosis, and it is not the purpose of this chapter to restate it. The purpose of this chapter is to ask what governance reform can actually *do* about each of these conditions, because for every one of them there is a policy lever that has either been left unpulled or pulled in a way that works against itself.

Consider the conditions in turn, not to re-describe them but to locate the reform that addresses them. Aging infrastructure persists because the economics of replacement are prohibitive and because no policy mechanism aligns the cost of modernization with the national interest in resilience; governance reform addresses this not by ordering rip-and-replace, which is a multi-decade impossibility, but by reforming procurement and incentive structures so that the long, unavoidable transition is steered toward secure-capable equipment and away from the next generation of technical debt. Insecure-by-design systems persist because the broader market has largely failed to make security a condition of sale; governance reform addresses this through secure-by-design regulation, shifting the burden of safety from the operator who inherits a defective product to the manufacturer who ships it. The absence of standardization persists because no body has the mandate, or the convening authority, to harmonize the proprietary chaos; governance reform addresses this through standards harmonization and through the strategic use of government purchasing to make common, interoperable, secure baselines a market expectation rather than a regulatory wish. And deep interdependency persists, in governance terms, because the agencies and sectoral bodies responsible for the interlocking systems are themselves siloed, with overlapping and uncoordinated mandates; reform addresses this by streamlining those mandates so that the cross-sector nature of the risk is matched by a cross-sector coherence of authority. Each technical condition, in other words, has a governance counterpart, and it is the governance counterpart that this chapter is built to treat.

The point of this reframing is not rhetorical. It is to resist the seduction of the technical fix as a substitute for the structural one. A water utility cannot patch its way out of a flat network it cannot afford to re-architect; but a procurement reform that makes network segmentation a baseline requirement for federally supported infrastructure changes the calculus for every utility that follows. A manufacturer will not voluntarily abandon hardcoded passwords as long as the cost of doing so is borne by the manufacturer and the cost of leaving them is borne by the customer; but a secure-by-design mandate relocates that cost to where it belongs. True resilience, as Chapter 8 argued and Chapter 13 will develop, is built not only by repelling attacks but by strengthening the underlying structure of the systems, and the lever that strengthens structure at national scale is not exhortation but reform of the rules that govern investment, procurement, and design.

Streamlining and Harmonizing Overlapping Mandates

The web of federal agencies tasked with safeguarding the nation's operational technology, while well-intentioned, can itself become a source of vulnerability. Chapter 3 mapped that web in detail, the crowded federal landscape in which the Cybersecurity and Infrastructure Security Agency, the Department of Energy, the National Security Agency, the Department of Defense, the Federal Bureau of Investigation, and a constellation of sector-specific regulators all hold authorities that touch industrial control systems, and traced how the resulting overlap produces friction, jurisdictional ambiguity, and a diffusion of accountability that no single agency is positioned to resolve. This chapter does not re-litigate that map. It asks the next question: given a landscape of overlapping mandates, what does reform actually consist of? The answer is not the abolition of agencies or the creation of a monolith, but the disciplined, often unglamorous work of clarifying, deconflicting, and harmonizing what already exists.

The complexity Chapter 3 described arose not from a lack of dedicated effort but from the historical accretion of agency roles, the emergence of new threats faster than statutes could be rewritten, and the genuine difficulty of carving clean boundaries across a technological landscape that refuses to hold still. The lived consequence is concrete. When a threat actor targets a component within a power grid's supervisory control and data acquisition system, questions immediately arise: does the lead investigative role fall to the FBI, is CISA the primary agency for coordination and information dissemination, or should the Department of Energy, with its deep sector expertise, take precedence? This ambiguity produces duplicated effort, delayed communication, and the very real possibility that critical information falls through the seams as agencies negotiate the boundaries of their own jurisdictions. Reforming OT cybersecurity governance, at this level, means streamlining federal agency mandates to produce clarity, eliminate redundancy, and assign distinct, accountable roles.

The first mechanic of reform is the rigorous identification and elimination of redundant authorities. This requires a deep audit of the statutory authorities, programmatic activities, and operational directives of each relevant agency. It is not uncommon for multiple agencies to hold both the legal standing and the operational capacity to perform substantially similar functions, conducting vulnerability assessments, issuing cybersecurity guidance, responding to incidents. Some redundancy is deliberately engineered into critical systems for resilience, but in the realm of governance and policy execution it typically produces inefficiency and a loss of focused accountability. When both CISA and a sector-specific agency are empowered to issue overlapping directives to the same regulated entities, the result is conflicting guidance, confusion, and a dilution of the authoritative voice the regulated entity needs. The objective of streamlining is not to eliminate valuable functions but to ensure each function is performed by the most appropriate entity, or, where it must be shared, that lead and supporting roles are explicitly coordinated. The instruments of that coordination are familiar but underused: formal Memoranda of Understanding that define lead and supporting roles in writing, and, where memoranda cannot reach, legislative adjustment to clarify agency missions outright.

The second mechanic is the affirmative allocation of defined, non-overlapping roles based on core competency. As Chapter 3 noted, CISA has been designated the nation's central hub for cyber defense and risk management, which positions it to coordinate federal effort, share threat intelligence, and provide cross-sector guidance. But sector-specific agencies hold domain knowledge and established relationships that CISA cannot replicate; the Department of Energy's understanding of the energy sector's interdependencies, the FBI's investigative mandate, the NSA's intelligence and advanced-capability focus are distinct and complementary. Reform means recognizing and leveraging these specialized roles without permitting them to encroach on one another, so that a regulated organization knows, without ambiguity, whom to engage for a given form of support. This clarity is not bureaucratic tidiness for its own sake; it is the precondition for a predictable cybersecurity ecosystem in which guidance is coherent and response is fast.

The payoff of harmonization is most visible at the two points where the current fragmentation costs the most: policy implementation and incident response. When mandates are clear and non-overlapping, the development and dissemination of policy becomes coherent, and an organization can rely on a single authoritative source for the guidance relevant to its sector rather than reconciling a labyrinth of potentially conflicting directives. This reduces the administrative burden on the regulated and raises the likelihood that policy is understood, adopted, and actually implemented. During an incident the stakes of clarity rise sharply. A well-defined governance structure ensures that response is not hostage to jurisdictional dispute: if an attack on a water treatment facility raises public-health concerns, the agency with primary coordinating responsibility, whether CISA, the EPA, or another designated entity, takes the lead and integrates support from the FBI for investigation or FEMA for consequence management without the costly negotiation of who is in charge. Streamlined mandates convert a fragmented federal posture into an integrated one, ensuring that intelligence gathered by one agency flows to those who can act on it, and that responsibility for technical remediation, criminal investigation, and consequence management is assigned rather than contested.

This deconfliction problem does not stop at the water's edge, and here the reform agenda must confront a complication that the domestic picture alone understates. The same impulse that produced a crowded federal landscape has, in recent years, produced a crowded regulatory landscape, as overlapping disclosure and security mandates proliferate faster than any single body harmonizes them. The Securities and Exchange Commission's 2023 rule requiring public companies to disclose material cybersecurity incidents within four business days of determining materiality[1], and the European Union's NIS2 directive, with its expanded sweep across essential and important sectors and its own incident-reporting timelines[2], are each, on their own terms, serious efforts to force the cost of insecurity into the open. But for a multinational operator of industrial infrastructure, they arrive as yet another layer of obligations to reconcile, a domestic disclosure clock running against a separate European one, sector definitions that do not align, reporting thresholds that do not match. This is mandate proliferation of exactly the kind reform exists to rationalize. The streamlining impulse, in other words, cannot be purely inward-facing. A genuinely reformed governance structure treats the harmonization of its own overlapping disclosure and security mandates, and the alignment of those mandates with the major international regimes its regulated entities must already satisfy, as a first-order objective rather than an afterthought, because every hour an operator spends reconciling contradictory mandates is an hour not spent reducing the mundane vulnerabilities that actually imperil the system.

The difficulty of this reform should not be understated. It requires agencies to cede certain authorities, or to define shared responsibilities in ways that enhance overall effectiveness rather than protecting individual turf. It demands sustained dialogue among agencies, legislators, and industry. The available avenues, creating new entities, consolidating existing programs, legislatively realigning responsibilities, each carry their own friction. The goal is to move away from a model in which many agencies are vaguely responsible for OT cybersecurity toward one in which each holds a precisely defined, accountable mission within a cohesive national strategy. Without this reform, every other improvement, every standard, every procurement requirement, every secure-by-design rule, is implemented through a structure that blunts its force; with it, the structure begins to amplify rather than dissipate the nation's effort.

Reforming Procurement: Buying Security Into the Market

Of all the levers available to governance reform, procurement is the most underused and, arguably, the most powerful. Government is not merely a regulator of the technology market; it is among its largest customers, and the conditions it attaches to its own purchasing radiate outward through supply chains long after the contract is signed. A manufacturer that must build a secure product to win a federal energy contract often finds it uneconomical to maintain a separate insecure assembly line for everyone else; the secured product tends to become the product, and the baseline rises for buyers who never sat at the negotiating table. Procurement reform, properly conceived, is industrial policy by other means: it converts the state's spending into a market signal powerful enough to pull vendors toward security without the heavier machinery of direct regulation. This is precisely why it belongs at the center of this chapter's agenda rather than the periphery, because it addresses the conditions Chapter 8 named, aging infrastructure, insecure-by-design products, the absence of common standards, at the only moment when changing them is cheap: the moment of purchase.

The first reform is to make security a nonnegotiable criterion of acquisition rather than a discretionary preference. For decades, OT procurement has been governed by the familiar trinity of cost, functionality, and delivery schedule, with security treated, if at all, as a soft factor easily traded away under budget pressure. The consequence is the steady accretion of exactly the technical debt Chapter 8 catalogued: equipment bought because it was cheapest and available, deployed for a thirty-year lifespan, and incapable from the first day of supporting the encryption, authentication, or patching that a secure posture requires. Reform inverts this. It writes security requirements into the acquisition rules themselves, so that a product incapable of meeting a defined baseline is simply ineligible, not merely disfavored. The baseline need not be exotic. It is the unglamorous, eminently fixable set of properties whose absence Chapter 8 identified as the real source of risk: the capacity for unique, changeable credentials rather than hard-coded defaults; support for authenticated and encrypted communication rather than the bare legacy protocols; the ability to be segmented rather than an architecture that presumes a flat, trusting network; and the basic logging and monitoring hooks without which an operator cannot see its own environment.

The second reform extends procurement's reach across the full lifecycle of the asset, because OT's defining characteristic, the one that makes its vulnerabilities so stubborn, is precisely the length of that lifecycle. A device bought today may still be controlling a pump or a turbine twenty-five years from now. Procurement reform that ends at the moment of delivery guarantees that whatever security the device shipped with will decay into obsolescence while the device itself remains in service. Lifecycle-aware procurement therefore makes ongoing security a contractual obligation of the vendor, not a hope of the operator. It requires committed support windows during which the manufacturer will furnish security updates; it requires a contractual duty of timely vulnerability disclosure and remediation rather than the silence that leaves operators blind; and it requires that the means of patching be built in, so that updating a device does not demand the costly physical intervention or operational shutdown that, in practice, means it never happens. The aim is to break the cycle Chapter 8 described, in which equipment outlives its security by decades, by binding the vendor to the asset's whole working life rather than to the instant of sale.

The third reform attacks the supply chain itself, which has become one of the principal vectors by which mundane compromise enters otherwise defended environments. A single exploited third-party component can nullify an operator's defenses entirely, and the operator often has no visibility into what its own equipment is built from. Procurement is the natural point of leverage. Reform here means requiring suppliers to furnish a software bill of materials, a verifiable inventory of the components and dependencies embedded in a product, so that when a vulnerability surfaces in a widely used library the operators who depend on it can be identified and warned rather than left to discover their exposure in the breach. It means extending security requirements contractually down the supply chain, so that a prime contractor cannot satisfy its obligations by integrating insecure subcomponents. And it means provenance and integrity assurances sufficient to give the buyer confidence that what was ordered is what was delivered, uncompromised in transit or at the point of manufacture. The Sector Coordinating Councils discussed later in this chapter have a natural role here: precisely because the vast majority of critical infrastructure is privately owned, an SCC can develop a standardized set of cybersecurity requirements for OT vendors that all member organizations adopt, aggregating private purchasing power into a market signal that reinforces, and extends beyond, the government's own.

The fourth and most strategic reform is to wield aggregate purchasing power deliberately, as an instrument of standards harmonization. The heterogeneity Chapter 8 identified, the patchwork of proprietary vendors, each with its own protocols and configurations, that makes consistent security nearly impossible, is in large part a market failure that procurement can help correct. When government and large private buyers converge on a common set of security baselines and interoperability expectations, the proprietary chaos begins to yield to de facto standardization, not because a regulator commanded it but because the market will not reward the vendor who refuses to meet the baseline its largest customers demand. This is the mechanism by which procurement reform and standards harmonization, treated separately in this chapter for clarity, in practice reinforce one another: harmonized standards give procurement a baseline to require, and coordinated procurement gives harmonized standards the market traction that voluntary frameworks, on their own, have rarely achieved at scale.

The difficulty of procurement reform is real and worth naming. Security-capable equipment frequently carries a higher acquisition cost, and budget cycles reward the cheaper line item against a risk that has not yet materialized. Smaller operators and resource-constrained municipalities, the water districts and rural utilities that figure so heavily in the OT risk picture, may struggle to meet elevated requirements without support. A reform agenda that ignores this reality will founder, which is why procurement reform must travel with the incentive alignment discussed below: grant conditions, cooperative agreements, and cost-sharing arrangements that allow the requirement to be met rather than merely imposed, and tiered expectations that scale to the capacity of the buyer. But the underlying logic is unassailable. The state spends enormous sums on the technology that runs the nation's critical infrastructure. Spending it without security conditions is to pay, twice, for the next decade of vulnerability; spending it with them is to convert routine purchasing into one of the most effective security interventions available to governance at all.

Secure-by-Design Regulation: Relocating the Burden of Safety

If procurement reform changes what the government and large buyers demand, secure-by-design regulation changes what every manufacturer is permitted to ship in the first place. It is the reform most directly aimed at the condition Chapter 8 named insecure-by-design, and it rests on a single, clarifying premise: that the security of a product is the responsibility of the entity that designs and builds it, not of the operator who inherits its defects. For most of the history of OT, that premise has been inverted. Manufacturers have shipped systems with default and hardcoded passwords, unauthenticated protocols, and architectures that trust all internal traffic, and the entire burden of compensating for these defects, segmenting the network, wrapping the protocol, monitoring the credential that cannot be changed, has fallen on the operator, who is least able to fix the root cause and most exposed to its consequences. Secure-by-design regulation relocates that burden to its source.

The principle is simple to state and consequential to enforce: security should be a foundational property designed into a product from its inception, not a feature added under duress after deployment or, worse, after a breach. In practice this means a manufacturer should not be able to lawfully sell into critical infrastructure a product that lacks the elementary properties whose absence the realized incidents, not the hypothetical ones, have shown to matter. The litany is by now familiar from Chapter 8, and the point of secure-by-design regulation is to make each item on it a manufacturing obligation rather than an operator's problem. Default and hardcoded credentials, the single condition behind the open-internet PLC with the password "1111," should be prohibited in favor of mechanisms that compel a unique credential before the device will operate. Unauthenticated and unencrypted communication, the legacy-protocol condition that lets any device on a network issue commands to a pump or a chemical feeder, should give way to authenticated, encrypted communication as the shipped default. Architectures that presume a flat, trusting network should yield to products that support segmentation and least privilege. And the basic instrumentation an operator needs to see its own environment, logging, monitoring, the capacity to detect anomalous behavior, should be built in rather than absent. None of this is exotic. It is, as Chapter 8 insisted, unglamorously fixable, and secure-by-design regulation is the mechanism that requires it to be fixed at the only point where fixing it is cheap.

A complementary principle, sometimes called secure-by-default, governs the state in which a product arrives. It is not enough that a system be *capable* of secure configuration if it ships in an insecure one and relies on an overstretched operator to harden it. Reform requires that the out-of-box state be the secure state: unnecessary services and ports disabled, the most protective settings enabled by default, and any departure from that posture a deliberate, documented choice by the operator rather than an accident of a permissive factory default. This matters acutely in OT, where the operator is frequently a small utility or a plant engineer without dedicated security staff, and where the gap between a product's secure capabilities and its insecure default is exactly the gap through which mundane compromise enters.

Secure-by-design regulation also reaches the manufacturer's conduct after the sale, and this is where it connects to the disclosure regimes that reform must rationalize. A secure development obligation is hollow if it ends at shipment, because vulnerabilities will be discovered in even well-built products across the long OT lifecycle. Regulation should therefore impose on manufacturers a duty of coordinated vulnerability disclosure and timely remediation: an obligation to receive and act on vulnerability reports, to issue fixes within defined windows, and to inform affected operators rather than leaving them to learn of their exposure from an attacker. Here the proliferation of disclosure mandates discussed earlier becomes directly relevant. The SEC's 2023 rule requiring disclosure within four business days of a materiality determination[3] obliges public companies to disclose material incidents to investors; NIS2[4] imposes its own incident-reporting duties across European essential and important entities. These are disclosure obligations that fire *after* an incident has occurred. Secure-by-design regulation works upstream of them, reducing the incidents that trigger disclosure in the first place, and a coherent reform agenda treats the two as complementary halves of a single design rather than as unrelated mandates accumulating in isolation: upstream rules that prevent the defect, downstream rules that surface its consequences, harmonized so that a manufacturer's disclosure duty to operators and an operator's disclosure duty to regulators and investors form a continuous chain of accountability rather than a set of disconnected clocks.

The hardest question secure-by-design regulation must answer is the question of liability, and reform cannot dodge it. As long as the manufacturer bears no consequence for shipping an insecure product and the operator bears the full cost of the resulting breach, the incentive to cut security corners at the design stage remains intact regardless of how the product is marketed. A serious reform agenda contemplates shifting a measure of liability toward manufacturers who ship products falling below a defined standard of care, so that the economic weight of insecurity is felt by the party with the power to design it out. This is delicate territory. Liability that is too blunt risks chilling the supply of equipment or driving smaller manufacturers from the market; a safe harbor for manufacturers who demonstrably adhere to recognized secure-development practices is the conventional and sensible counterweight, rewarding the manufacturer who builds securely with protection from open-ended exposure. But the direction of reform is clear. The current allocation, in which the operator absorbs the cost of the manufacturer's design choices, is the precise market failure that perpetuates insecure-by-design systems, and only by relocating that cost does regulation change the manufacturer's calculus at the drawing board.

Secure-by-design regulation and procurement reform are, in the end, two faces of the same intervention, and reform should deploy them in concert. Procurement requires security of what the government and large buyers purchase; secure-by-design regulation requires it of what any manufacturer may sell. The former works through the market and reaches as far as purchasing power extends; the latter works through law and reaches the whole market, including the resource-constrained operators whom no procurement requirement protects because they were never large enough to set terms. Together they address the insecure-by-design condition from both ends, converting it from a permanent feature of the OT landscape into a defect that the market is structured to drive out, which is precisely the kind of structural remedy that exhortation and the technical patch, on their own, have not been able to deliver.

Harmonizing Standards and Setting a National Baseline

The lack of standardization that Chapter 8 identified as a core vulnerability has a governance remedy distinct from, though entwined with, the procurement and secure-by-design reforms already discussed. That remedy is the deliberate harmonization of standards and the establishment of a coherent national baseline against which security across critical sectors can be measured. The need arises directly from the heterogeneity of the OT world: a landscape so fractured across proprietary vendors, sector-specific frameworks, and inconsistent configurations that consistent security policy, uniform monitoring, and standardized incident response are exceedingly difficult to achieve. Each new vendor and each sectoral silo introduces its own protocols, its own assumptions, and its own learning curve, and the cumulative effect is an environment in which organizations struggle to maintain even a clear understanding of their own risk posture.

The reform mechanic here is not the imposition of a single rigid standard dictating every technical control, which would be both impractical across the diversity of OT and counterproductive to the operational realities of distinct sectors. It is, rather, the development of an overarching national framework that establishes fundamental principles, risk-management expectations, and performance objectives that all critical sectors must meet, while leaving the sector-specific translation of those objectives to the bodies best equipped to perform it. Such a framework, built on existing standards and accumulated best practice rather than invented from nothing, provides a common language and a shared understanding of risk management across industries that today speak in mutually unintelligible dialects. It can define expectations for asset inventory, for vulnerability management, for incident response, and for supply-chain security, each tailored to the realities of OT rather than imported wholesale from the IT world where so many OT security failures originate in the misfit.

Standards harmonization is also the connective tissue between the domestic reform agenda and the international regulatory landscape that the nation's critical-infrastructure operators must already navigate. A reformed national framework should be developed in deliberate alignment with the major international regimes, engaging international standards bodies and coordinating with allied nations, so that the efforts of multinational operators are harmonized rather than multiplied. This is the same rationalization impulse that animates the streamlining of overlapping mandates, applied to the proliferation of standards: where the SEC's disclosure rule and the European Union's NIS2 directive[5] impose their own expectations on entities that often operate across both jurisdictions, a harmonized national baseline that maps cleanly to recognized international standards reduces the friction of compliance and frees operator attention for the substance of security rather than its paperwork. Harmonization, in this sense, is not merely domestic tidiness; it is the alignment of the nation's baseline with the regimes its own regulated entities cannot escape.

Crucially, a harmonized standard is the prerequisite that makes the other reforms in this chapter operable. Procurement cannot require a baseline that does not exist; secure-by-design regulation needs a recognized standard of care against which a manufacturer's conduct can be judged and a safe harbor defined; performance metrics need a common yardstick against which improvement can be measured across sectors that otherwise report in incommensurable terms. The development of such a framework must, of course, proceed through extensive consultation with sector-specific regulators, industry stakeholders, and technical experts, both to ensure that the resulting baseline is practical and implementable and to secure the buy-in without which any standard becomes a document rather than a practice. But the strategic point is that standards harmonization is not one reform among several independent reforms; it is the foundation on which procurement reform, secure-by-design regulation, and accountability all rest, the common reference frame that turns a collection of separate interventions into a coherent governance architecture.

Empowering Sector Coordinating Councils (SCCs)

Streamlining federal mandates, reforming procurement, regulating for secure design, and harmonizing standards are all, in the first instance, exercises of public authority. But the vast majority of critical infrastructure is privately owned and operated, and a reform agenda confined to government action alone cannot fully reach it. The structures that bridge the public and private sectors are therefore not a supplement to the reform agenda but an integral part of it, and among these the Sector Coordinating Councils stand out. These councils, typically comprising industry representatives within a specific critical-infrastructure sector alongside relevant government agencies, are uniquely positioned to foster information sharing, coordinate risk-management strategy, and drive the adoption of the very baselines and secure-procurement practices the preceding reforms establish. Yet their effectiveness is routinely constrained by limited resources, unclear mandates, and inadequate access to timely, actionable intelligence. Empowering them is the reform that translates public policy into private practice on the ground.

The fundamental role of an SCC is to operate the public-private partnership within its sector, and that partnership is indispensable precisely because of the ownership structure just noted. Government agencies hold broad threat intelligence and national strategic insight but lack granular operational understanding of specific industrial environments; industry holds deep knowledge of its own systems, constraints, and the practical difficulty of implementing security measures but lacks the national vantage. SCCs are the nexus where these perspectives converge, the formal mechanism through which industry articulates its needs and constraints to government and government conveys its threat assessments and policy requirements to industry in a form that is relevant and actionable. This bidirectional flow is what keeps the reforms of this chapter from becoming abstractions: an energy-sector SCC can surface, to the Department of Energy, the concrete vulnerabilities of advanced metering infrastructure or distributed energy resources, and government can route its requirements back through the same channel in terms a utility engineer can act on.

Empowering SCCs to play this role requires, first, that their mandates be clearly defined and formally recognized within the national framework rather than left to informal collaboration. CISA and the sector-specific agencies should delineate concrete responsibilities for SCCs, in threat-intelligence sharing, in incident-response coordination, in the development of sector-specific standards and guidelines that implement the national baseline, so that SCCs are recognized as integral components of national resilience rather than merely consultative bodies. A clearly defined mandate might task an SCC with collecting and disseminating anonymized incident data from its sector to CISA, which can then aggregate it into sector-wide trend analysis and national threat assessment. This formalization confers the legitimacy and authority that encourages member organizations to participate and commit.

Empowerment requires, second, enhanced resources, because mandates without means are an unfunded aspiration. Many SCCs operate today on volunteer effort and minimal budgets that cannot support the strategic, sector-wide work now being asked of them. Financial support, whether direct or through grants and cooperative agreements, can fund dedicated staff, advanced information-sharing platforms, training, and research. Technical resources and expertise from government, access to government-developed tools, training in threat-analysis methodologies, dedicated agency liaisons, can materially raise an SCC's capability; CISA might, for instance, extend access to its threat-intelligence platforms or train SCC members in conducting supply-chain risk assessments for OT components, the very assessments that lifecycle-aware procurement reform requires.

Empowerment requires, third, improved access to government information, which is in many sectors the binding constraint. SCCs need timely, accurate intelligence about emerging threats, vulnerabilities, and attack methodologies targeting their sector's OT, and they need it in a form digestible and directly applicable to operational reality. The current dissemination process is too often slow, over-classified, or too general to be of practical value to an industrial operator. Empowering SCCs means establishing streamlined, secure channels, dedicated information-sharing environments or trusted information networks, through which actionable intelligence can move quickly. When intelligence indicates a new exploit targeting a programmable logic controller common in manufacturing, the relevant agency should work through the manufacturing SCC to push that warning, with recommended mitigations, to every member company before the exploit is widely used rather than after.

Beyond these enabling conditions, SCCs serve as force multipliers for the reform agenda in ways government cannot replicate directly. They can foster a culture of security within their sectors through workshops, awareness campaigns, and peer-to-peer learning, reaching even the smallest and least sophisticated organizations, an effect that matters disproportionately for the small and medium-sized operators who lack the resources to confront OT security alone. They can aggregate demand for security services and shared resources. And, as noted in the discussion of procurement, an SCC can develop a standardized set of cybersecurity requirements for OT vendors that all member utilities adopt, extending secure-procurement discipline across a sector's supply chain through private coordination rather than public mandate. This is the federated model of Chapter 9 made operational: not government commanding industry, nor industry left to its own devices, but the two working through strengthened intermediary structures that carry national policy into sectoral practice and sectoral knowledge back into national policy.

Aligning Incentives and Establishing Accountability

The reforms assembled in this chapter share a common dependency: whatever clarity of mandate, rigor of procurement, stringency of secure-by-design rule, or coherence of standard reform achieves, its ultimate efficacy hinges on whether the incentives surrounding the regulated entities pull toward security or away from it, and on whether there exist accountability mechanisms robust enough to keep well-crafted policy from decaying into bureaucratic inertia. Aligning incentives is therefore not a separate item on the agenda but the connective logic that determines whether the rest of it works, and it must be designed in deliberately rather than assumed.

Measuring the success of a reformed governance model, one that relies on the dynamic interaction between federal agencies and a privately owned critical infrastructure, demands a more nuanced approach than the compliance-based indicators on which oversight has traditionally leaned. Adherence to regulations and reporting requirements has its place, but it captures process rather than outcome, and the outcome that matters is the actual reduction of risk and improvement of resilience. Performance metrics should be built to capture that outcome. For federal agencies, this means measuring the speed and accuracy of threat-intelligence dissemination to the SCCs, the demonstrable effect of vulnerability assessments on reducing exposure, and the effectiveness of inter-agency coordination during incidents, a direct check on whether the mandate-streamlining reform discussed earlier is producing the coherence it promises. A metric for CISA might be the average time to deliver actionable intelligence to an SCC after a significant OT threat emerges, or the proportion of critical-infrastructure organizations that demonstrably implement recommended controls following a sector-specific alert. The Department of Energy might be measured on the adoption rate of advanced OT security practices within the energy sector; the FBI on the disruption of adversary campaigns against OT, the attribution of significant attacks, and resulting prosecutions. For the SCCs, metrics should reflect their bridging role: the volume and quality of information flowing in both directions, the successful implementation of sector-wide initiatives developed with government partners, the measurable maturation of member organizations, and the rate at which sector-specific standards implementing the national baseline are actually adopted.

Accountability mechanisms must then ensure these metrics are pursued rather than admired, with genuine consequences for failure. For federal agencies, accountability can run through existing oversight, Inspector General reports, congressional hearings, performance reviews tied to budgets and leadership evaluation, adapted to address OT cybersecurity specifically. Regular audits of threat-intelligence-sharing and incident-response coordination, with findings publicly reported, give the streamlining reform teeth; agency leaders should answer for failures of inter-agency collaboration or for delay in responding to critical OT threats within their purview, and the clarity of mandate the reform establishes should itself become a measure of accountability, so that an agency failing to perform its defined, non-overlapping role bears that failure in its assessment.

For the private sector, accountability is more delicate, because it must balance regulatory enforcement against incentivized self-improvement, and here the incentive-alignment logic comes fully into view. Direct mandates and penalties are necessary for baseline compliance, but the more effective path to genuine resilience runs through incentive rather than penalty. This is where the procurement and accountability reforms converge: a tiered system of engagement can reward organizations that demonstrate higher OT cybersecurity maturity and actively participate in collaborative structures like the SCCs with concrete benefits, streamlined regulatory processes, access to advanced threat intelligence, preferential treatment in government contracts, the last of which closes the loop directly with procurement reform by making security maturity a competitive advantage in winning the government's business. Conversely, persistent failure to address identified vulnerabilities that pose significant risk to national security or public safety should trigger escalating consequences, increased regulatory scrutiny, mandatory remediation plans, and, in the extreme, fines or operational restrictions. The architecture is one of carrots reinforcing the reform agenda and sticks reserved for the negligence that the agenda exists to prevent.

The SCCs play a distinctive role in this accountability structure, serving as internal accountability bodies for their sectors. By developing and promoting sector-specific best practices and maturity models, they let members benchmark their performance, generating the peer pressure and shared commitment that external regulation cannot manufacture; and by aggregating and reporting sector-wide performance data to government, they provide the consolidated view of progress that informs national policy and resource allocation. Their effectiveness as accountability mechanisms depends, in turn, on their own governance and transparency, clear internal procedures, fair representation across member types, and honest reporting of activities and outcomes, without which the accountability they are meant to supply becomes circular.

A central challenge runs through all of this: the inherent difficulty of attributing specific security outcomes to specific actions in an environment as complex and interconnected as OT. A sophisticated attack involves multiple vectors, and its success or failure turns on a confluence of human, technological, and defensive factors that resist clean causal assignment. Metrics must therefore be designed to capture trends and leading indicators of improvement rather than a simplistic cause-and-effect ledger. Rather than the unmeasurable "number of attacks prevented," a realistic metric tracks the reduction in time to detect and respond to a simulated OT intrusion, or the increase in the use of multi-factor authentication on OT networks, indicators that speak to the strengthening of defensive capability and response readiness, which are the precursors to prevention and the proper objects of a reform agenda that cannot promise to abolish risk but can promise to reduce it measurably.

The notion of accountability must itself be defined with care. In the OT context it should not collapse into mere punishment for failure, but should encompass responsibility for continuous improvement and risk management. The aim is to foster an environment in which organizations learn from incidents, share lessons without fear of undue reprisal, and invest proactively in their posture, accountability framed as a commitment to ongoing learning and adaptation, driven by measurable progress and transparent oversight. An organization that suffers a significant OT incident is properly held accountable for conducting a thorough post-incident analysis, sharing findings with its SCC and the relevant agencies, and implementing a remediation and prevention plan whose success, measured against improved security metrics, demonstrates accountability in the constructive sense. To anchor this, government should work with the SCCs and industry to articulate clearly what constitutes a "critical failure" in OT security, the situations, disruptions to power, water, or healthcare that pose an immediate threat to public health and safety, severe enough to warrant strong, coordinated response and, where negligence is established, regulatory intervention. Defining these thresholds makes performance expectations specific and accountability enforceable when the essential safety nets are compromised through a failure of due diligence.

None of this can be static. The OT threat landscape evolves continuously, vulnerabilities emerge, adversary techniques advance, and the metrics and accountability mechanisms must evolve with it. Regular review of existing metrics, ongoing engagement with stakeholders to identify new indicators of progress, and periodic adjustment of accountability frameworks are necessary to keep the reforms of this chapter relevant and effective over time. This iterative cycle of measurement, feedback, and adaptation is the mechanism by which a reform agenda stays alive rather than ossifying into the very inertia it was meant to dispel.

The reforms gathered in this chapter, streamlining and harmonizing overlapping mandates, reforming procurement, regulating for secure design, harmonizing standards, empowering the structures that reach private industry, and aligning the incentives that govern them all, are best understood not as a menu of independent options but as a single interlocking architecture. Standards harmonization gives procurement and secure-by-design regulation a baseline to require; procurement and secure-by-design regulation give that baseline market and legal force; streamlined mandates ensure the whole apparatus is administered through a coherent rather than a fragmented federal structure; the SCCs carry it into the privately owned infrastructure that constitutes most of the target; and aligned incentives with robust accountability determine whether any of it changes behavior on the ground. The success of this reform agenda will ultimately be judged, as Chapter 8's diagnosis implied and Chapter 13's resilience playbook will assume, not by the policies enacted or the structures erected, but by their tangible effect on the security and resilience of the nation's critical infrastructure, an effect that can be assessed only through rigorous measurement and unwavering accountability, and achieved only by treating the systemic vulnerabilities of OT as what they finally are: not an unsolvable technical mystery, but a structural condition that the right reform of the right rules can, deliberately and over time, repair.

Chapter 12

THE DELICATE BALANCE: STATE AUTHORITY, INDUSTRY, AND CIVIL OVERSIGHT

Securing Operational Technology (OT) in critical infrastructure requires a clear line between state authority and private sector responsibility. This balance is essential to protect national security without infringing on the operations of private asset owners. Effective governance must be both protective and respectful of the roles inherent in a market economy.

The state's interest in OT cybersecurity stems from its duty to protect national security, public safety, and economic stability. Disruptions to power grids, water plants, or transportation networks threaten national resilience, giving the government a clear mandate to establish baseline security requirements and intervene when systemic risks emerge. This authority includes setting standards, requiring incident reporting, and conducting oversight.

The Boundaries of State Authority

This authority is not unlimited. Private asset owners are driven by market forces and possess deep, specialized knowledge of their own systems. Overly prescriptive government mandates can stifle innovation, misallocate resources, and introduce new vulnerabilities. For instance, mandating a specific security software for all oil pipelines might be technologically outdated upon implementation or unsuitable for different operational environments. The state's role is to mandate outcomes, not methods. The 'how' should remain with the asset owner.

This tension is clear in cybersecurity standards. While agencies such as CISA, DOE, and the EPA — each operating under different statutory authorities — develop guidelines and recommend frameworks such as the NIST Cybersecurity Framework, mandating specific technical controls across diverse sectors is problematic. The risk is that such mandates become technologically myopic and create compliance burdens that outweigh their security benefits. A more effective approach is for the state to set performance goals, e.g., "reduce the likelihood of unauthorized control of critical industrial processes by X percent", and allow industry to determine the most effective and efficient technical and procedural means to achieve these goals. This approach respects industry expertise and fosters ownership over security outcomes.

Furthermore, the state's authority should be exercised with an understanding of the economic impact on asset owners. Implementing robust cybersecurity measures, particularly in aging OT environments, can be costly and complex. Therefore, state mandates must be proportional, feasible, and justified by clear risks to avoid imposing undue financial burdens that can lead to resistance or unsafe workarounds. This could involve phased timelines, financial incentives for adopting new security measures, or technical assistance, especially for smaller operators.

The definition of "critical infrastructure" also affects the boundaries of state authority. While there is broad consensus on core sectors like energy, water, and defense, the digital transformation has blurred the lines, incorporating formerly less critical sectors into interconnected supply chains that are now essential. As these sectors become more integrated, the rationale for government oversight expands, but the nature and extent of that oversight must adapt. A small, regional internet service provider might be deemed critical due to its role in enabling communication for other critical sectors, but the specific cybersecurity requirements imposed on it should differ significantly from those placed on a national power grid operator. State authority should be granular, with requirements tailored to the risks and impact profiles of different entities.

There is a further boundary the fear narrative tends to erase altogether: the civil one. Every invocation of imminent catastrophe is also, implicitly, an argument for expanded state reach — for broader monitoring of privately owned networks, for emergency authorities invoked outside ordinary deliberation, for information-sharing regimes that draw operational and sometimes personal data into government hands. Some of this is justified; some of it is necessary. But fear is corrosive to the scrutiny that should accompany any growth of surveillance power, because to object to a measure marketed as protection against derailed chemical trains is easily recast as recklessness. The predictable result is ratchet-like: monitoring authorities accrete during moments of alarm and seldom recede once the alarm passes, hardening into a standing apparatus whose proportionality no one is any longer incentivized to revisit. A governance model worthy of the name treats the protection of civil liberties not as an obstacle to security but as one of its measures. A surveillance authority that cannot survive sober, unfrightened review is not a security gain; it is a deferred liability. The same evidentiary discipline this book demands of threat claims must apply, in equal measure, to the powers asserted in their name.

Incident reporting requires careful calibration. The state needs timely incident data to understand the threat landscape, but mandatory reporting can be seen as burdensome. Asset owners may fear punitive action, reputational damage, or the disclosure of sensitive information. Effective reporting requires clear thresholds, a standard definition of a reportable incident, and assurances of confidentiality. The reporting mechanism should also be unobtrusive, ideally integrated with existing operational processes rather than creating a separate bureaucratic hurdle.

Mandating government penetration testing for OT systems presents similar challenges. While proactive testing is invaluable, mandating specific methods is problematic because asset owners know their systems' vulnerabilities and often conduct their own testing. Government-mandated testing, if not carefully designed, could lead to duplicative efforts, operational disruptions, or the revelation of sensitive security strategies. A better approach is for the state to set a performance objective for vulnerability management and incentivize asset owners to conduct regular, rigorous testing, perhaps by leveraging industry-led initiatives. In some high-risk scenarios, however, direct government authority to conduct or oversee such assessments might be justifiable if a clear and present danger is identified.

Centralized Defense: The USCYBERCOM "Managed Defense" Concept

One proposed evolution in public-private defense is a "managed defense" model in which U.S. Cyber Command (USCYBERCOM) would, under carefully defined authorities, partner directly with critical infrastructure owners. This conceptual approach moves beyond traditional information sharing to place government cyber operators, with explicit consent, inside private networks. Their mission would be to actively hunt for advanced persistent threat (APT) actors, applying national defense capabilities to protect privately owned assets vital to national security. This model is sometimes described as a domestic analogue to USCYBERCOM's overseas "hunt forward" operations, represents a proposed new frontier in collaborative defense, but any implementation would require careful navigation of legal authorities — including Title 10 authorities and the contested applicability of Posse Comitatus Act restrictions to domestic military cyber activity — and strict adherence to privacy safeguards.

Scope, Authorities, and Selection Criteria The managed defense concept envisions not a blanket surveillance program but a voluntary, mission-based engagement. Its proposed legal authority would be derived from USCYBERCOM's Title 10 authorities and its assigned mission, under the Unified Command Plan, to defend Department of Defense networks and, when directed, to conduct cyberspace operations in defense of the nation. Applying this authority domestically, however, would require a framework built on the private partner's explicit, informed consent, which would serve as the cornerstone of any such program's legality.

Participation would be highly selective and targeted. The selection criteria for pilot partners prioritize mature organizations in high-risk sectors, such as energy, defense industrial base, or finance, that face sophisticated nation-state threats. A candidate organization must demonstrate high cybersecurity maturity, including a dedicated security team capable of collaborating with government operators. This ensures the partnership is a collaboration, not a rescue mission for an unprepared entity.

The scope of each engagement is narrowly defined. The primary activities include: **Threat Hunting:** Proactively searching for signs of adversary presence, tools, and infrastructure that may have evaded the company's existing defenses. **Monitoring Support:** Providing advanced monitoring capabilities and analytics to detect sophisticated threats in real time. **Incident Support:** Offering expert assistance in analyzing and responding to a major incident, helping to contain the damage and eradicate the adversary.

Privacy Guardrails and Operational Controls To address significant privacy and civil liberties concerns, such engagements would be governed by a robust set of guardrails and operational controls:

- **Explicit Consent and Partner Control:** The private asset owner provides explicit, written consent before any activity begins. The company maintains full control over its networks and can halt the operation at any time. The company's security team works directly alongside the government hunters, ensuring full visibility into their actions.

- **Strict Rules of Engagement (ROE):** A detailed ROE is negotiated and agreed upon before the mission starts. This document specifies exactly which systems can be accessed, what tools can be used, what data can be collected, and how that data will be handled. The focus is on identifying adversary activity, not collecting business intelligence or personal data.

- **Data Minimization and Ownership:** Data collection is minimized to only what is absolutely necessary for the threat-hunting mission. Any data collected remains the property of the company, and government teams operate under strict handling protocols to protect sensitive information.

- **Layered Oversight:** The operations are subject to multiple layers of review, including by Department of Defense lawyers, civilian leadership within the Department of Defense, and relevant congressional committees. Information shared with other agencies for law enforcement or intelligence purposes must follow established legal channels.

Termination and "Stop-Rules" Clear "stop-rules" are essential to prevent mission creep and ensure the private partner remains in control. The engagement would automatically terminate under several conditions: upon the partner's request, at the end of a predefined operational period, or once the mission's objectives have been met.

If the government team discovers activity outside the agreed-upon ROE, they must immediately stop, report the finding to the company and their chain of command, and seek new authorization before proceeding. This mechanism ensures that the mission remains tightly focused and respects the legal and operational boundaries established at the outset. This pilot model tests a new form of collaboration, embedding national expertise on the front lines within a framework of trust grounded in private ownership, legal oversight, and mutual respect.

Industry as the Frontline Defender

Because critical infrastructure in the United States is largely (though not uniformly) privately owned, national cybersecurity strategy depends on industry's active participation. The focus must be on how to incentivize, empower, and collaborate with industry to foster proactive cybersecurity investment.

Incentivizing proactive investment requires a mix of financial support and clear liability frameworks. Tax credits, grants, or low-interest loans can help offset the high cost of security controls, especially for smaller operators. Clear liability protections for companies that demonstrate due diligence can also encourage investment beyond minimum compliance.

Empowering industry means providing adaptable tools and knowledge. Rather than prescribing rigid technical solutions, government should set performance-based goals, promoting flexible, risk-based approaches like the NIST Cybersecurity Framework. Government can also support the sharing of industry-specific threat intelligence and best practices through sector-based Information Sharing and Analysis Centers (ISACs), which are typically industry-led non-profits originally encouraged by Presidential Decision Directive 63 (1998) and subsequent critical-infrastructure policy.

Trust is the bedrock of this partnership, built on transparency, predictability, and a shared understanding of each party's constraints. The government must be transparent about its security objectives, the rationale behind its policies, and the potential risks associated with inaction. Industry, in turn, must be open about its operational challenges, technological limitations, and the realities of market pressures. This requires consistent and open channels of communication, moving beyond traditional, often adversarial, regulatory relationships. Regular dialogues, joint exercises, and collaborative working groups can help bridge the gap in understanding and foster a sense of shared purpose.

The concept of "due diligence" in cybersecurity is paramount here. When industry stakeholders can demonstrate that they have taken reasonable and appropriate steps to secure their systems, based on best practices and evolving threat landscapes, this can serve as a significant mitigating factor in the event of an incident. This requires a shift in government oversight from purely punitive measures to a more supportive approach. Rather than solely focusing on penalizing failures, governments can actively guide and assist industry in achieving higher security postures. This might involve offering technical assistance, providing access to specialized expertise, or developing training programs tailored to OT cybersecurity challenges.

Tangible examples of public-private partnerships include the collaborative development of sector-specific security guidelines. Organizations like the American Petroleum Institute (API) have developed voluntary industry standards, while the North American Electric Reliability Corporation (NERC), certified by FERC in 2006 as the Electric Reliability Organization under section 215 of the Federal Power Act, develops Critical Infrastructure Protection (CIP) standards for the bulk power system that become mandatory and enforceable once approved by FERC. These industry-led efforts leverage deep expertise to create practical security measures, and government support can amplify their impact.

Furthermore, the education and workforce development aspect cannot be overlooked. The cybersecurity talent gap is acute in the specialized field of OT security. Industry, in partnership with educational institutions and government agencies, plays a crucial role in developing training programs, apprenticeships, and certifications to build a skilled workforce capable of defending these complex systems. This includes not only technical skills but also an understanding of operational context and risk management.

Proprietary technology presents a unique challenge. Asset owners are protective of their system designs, so government agencies must handle information requests with sensitivity. To maintain cooperation, information sharing must be narrowly tailored to security concerns and data must be handled confidentially. This builds confidence that the government's focus is on collective security, not bureaucratic intrusion.

As cyber threats evolve, the government-industry relationship must be dynamic, with periodic policy reviews to adapt to new technologies. The goal is a resilient ecosystem that leverages both state authority and industry expertise.

The integration of OT security into broader corporate governance and risk management frameworks is another critical aspect of industry's role. Leading organizations are increasingly recognizing that cybersecurity is not solely an IT or OT department issue, but a fundamental business risk that requires board-level attention. This elevated focus encourages greater investment, better resource allocation, and a more holistic approach to security. Governments can encourage this by promoting awareness of OT cyber risks at the executive and board levels, perhaps through industry-specific risk briefings or the development of corporate governance guidance related to cybersecurity.

The Critical Role of Civil Oversight

This partnership is incomplete without civil society. Advocates, academics, and journalists scrutinize government and industry, holding them accountable.

Without this independent oversight, the delicate balance required for effective and equitable cybersecurity policy can easily tip, driven by the very real, but sometimes exaggerated, specter of cyber threats.

The concept of the "fear machine" is particularly pertinent here. In the high-stakes arena of national security, the consequences of failure create incentives to amplify threats. Governments may lean toward overreaching security measures, while industries might advocate for self-serving regulations. Civil society acts as a counterweight, injecting reason and evidence into debates that might otherwise be dominated by fear.

Academic institutions provide foundational oversight through rigorous, independent research. They deconstruct threats, evaluate security measures, and explore the ethical implications of policy, grounding public debate in empirical evidence rather than speculative anxiety.

Investigative journalists are also critical, examining opaque government and corporate operations to uncover negligence or overreach. By translating complex issues into accessible narratives, they empower the public to demand transparency and accountability.

Civil society organizations (CSOs), like digital rights groups, bring citizen perspectives into policymaking. They mobilize public opinion, lobby officials, and challenge policies that are inadequate or intrusive. For example, a CSO might raise concerns about the privacy implications of widespread government access to network data, even if that access is justified on national security grounds. They can also advocate for greater public involvement in the design and oversight of cybersecurity initiatives, ensuring that the needs and rights of the citizenry are not an afterthought. These groups often operate with limited resources, relying heavily on volunteer efforts and public donations, yet their impact can be significant in pressing for the "human element" and fundamental liberties to be considered in cybersecurity policymaking.

Independent oversight is also needed to bridge the information gap between the state, industry, and the public. Civil society advocates for transparency, from government surveillance to industry security practices, to ensure security measures are proportionate and legitimate.

The "fear machine" can also manifest in the realm of incident response. Following a significant cyber attack, there is a natural tendency towards immediate, often reactive, policy changes. Governments may feel pressured to demonstrate decisive action, and industries may seek to deflect blame. In this charged atmosphere, civil society can provide a vital moderating influence. By advocating for a measured and evidence-based approach to post-incident analysis, CSOs can help prevent the implementation of hasty and potentially ineffective security mandates. They can also ensure that the lessons learned from an incident are applied broadly, not just to the specific entity that was targeted, but to the broader ecosystem of critical infrastructure. This involves advocating for comprehensive post-incident reviews that are not solely focused on assigning blame but on identifying systemic weaknesses and recommending practical, sustainable improvements.

Furthermore, civil society's engagement is essential for ensuring that cybersecurity strategies are not solely focused on technological solutions but also address the human and societal dimensions of security. This includes advocating for robust privacy protections, promoting digital literacy and cybersecurity education for all citizens, and ensuring that the benefits and burdens of cybersecurity measures are distributed equitably. For instance, a policy that mandates extensive data collection for security purposes might disproportionately affect vulnerable populations or curtail legitimate freedoms of expression. Civil society groups can champion the cause of these groups and ensure that their rights are protected within the broader cybersecurity framework.

Legislative scrutiny is another indispensable pillar of oversight. Congressional hearings, committee reviews, and debates allow elected representatives to question officials, dissect policy, and push back against unchecked expansion of state power. This process, combined with independent commissions, helps institutionalize oversight by providing impartial assessments and fostering dialogue.

Ultimately, a healthy democratic society must balance security with civil liberties and public trust. This cannot be achieved by the state and industry alone. The scrutiny from civil society, academia, and the public is not an impediment to security but a precondition for its legitimacy. Oversight ensures that cybersecurity policies serve the public interest, are based on evidence, and are implemented in a way that respects democratic values. Without it, the danger of succumbing to the "fear machine" and creating restrictive or ineffective policies becomes a genuine risk to democratic governance.

Mechanisms for Collaboration

Robust OT cybersecurity requires an ecosystem of collaboration among government, industry, and civil society, built on concrete mechanisms for engagement. Structured dialogues are regular, formal forums that bring together key stakeholders and serve as a fundamental mechanism for collaboration. These dialogues can take various forms, from high-level summits involving cabinet-level officials and C-suite executives, to more specialized working groups focused on specific sectors or threat vectors within OT environments. For instance, a national dialogue on OT security could include representatives from key sectors, government agencies, academic researchers, and privacy advocates. Such forums provide a neutral ground for discussing threats, sharing best practices, and identifying policy gaps.

Joint working groups address specific challenges, such as developing industry-specific standards or creating a unified incident response playbook. These groups leverage the specialized expertise of their members: government provides regulatory context, industry brings operational knowledge, and civil society contributes perspectives on user impact and ethics. Working groups require balanced representation, clear mandates, and defined deliverables. Their success can be measured by their ability to produce actionable recommendations, pilot programs, or draft policies that have broad stakeholder support.

Transparent information-sharing forums are another critical component. Effective defense requires secure, trusted channels for sharing anonymized threat intelligence, vulnerability disclosures, and lessons learned from incidents. Government agencies can play a pivotal role in aggregating and disseminating threat information, often benefiting from classified intelligence that can be declassified and shared appropriately. Industry partners can share insights into the specific attack vectors and malware strains they are encountering in their OT environments. Civil society, particularly academic researchers, can contribute by identifying systemic vulnerabilities or emerging attack methodologies that might not be immediately apparent from individual incident reports.

Platforms for such sharing could include secure online portals, dedicated Information Sharing and Analysis Centers (ISACs) for OT, or secure communication channels established during collaborative exercises. The key principles are trust, timeliness, and relevance; information must be actionable and shared without compromising operational security or competitive advantage.

Collaborative research initiatives can also enhance OT cybersecurity. By pooling resources, government, industry, and academia can accelerate the development of new security technologies and policies. Jointly funded projects could focus on secure-by-design principles, advanced intrusion detection, or resilient supply chains. In this model, academia provides foundational research, industry offers real-world testing grounds, and government provides funding and policy direction. Initiatives should extend beyond technical solutions to explore human factors, training programs, and the ethics of new technologies. For example, a project might investigate effective training for OT operators or assess the operational impact of AI-driven anomaly detection.

Developing common taxonomies and terminology is another practical way to facilitate collaboration, as disparate language can create significant misunderstandings. For example, what one entity refers to as a "critical vulnerability," another might label a "significant threat." Establishing a shared lexicon for threats, vulnerabilities, incidents, and mitigation strategies can streamline communication and ensure that all parties are speaking the same language. This can be achieved through the development of standardized glossaries, common frameworks for risk assessment, and consistent reporting templates.

Dedicated liaison roles within government and industry can also enhance collaboration. These individuals act as primary points of contact, tasked with understanding and translating the needs of their respective sectors for their counterparts. This can help to bridge cultural divides and overcome organizational barriers that might otherwise impede effective cooperation. For instance, a government liaison officer could work closely with an industry ISAC to understand the specific challenges faced by its members, while an industry liaison within a government agency could ensure that regulatory proposals are informed by the practical realities of OT operations.

The concept of public-private partnerships (PPPs) can also be adapted and applied to the realm of OT cybersecurity. While often associated with infrastructure development, PPPs can be structured to focus on shared security objectives. This might involve jointly developing and deploying security technologies, sharing the costs and benefits of cybersecurity training programs, or collaborating on incident response exercises. The success of such partnerships relies on clear agreements outlining roles, responsibilities, risk-sharing, and intellectual property rights, ensuring that all parties are aligned and committed to the shared endeavor. Independent bodies, like national cybersecurity centers, are crucial for orchestrating these efforts. Operating with impartiality, they can act as conveners and knowledge brokers, creating a neutral environment for collaboration by conducting research, developing guidelines, and hosting forums. Their independence from the direct operational and regulatory pressures faced by government and industry allows them to foster innovation and build consensus more effectively.

Chapter 13

BEYOND FEAR: SHIFTING FROM PERCEIVED THREATS TO SYSTEMIC RESILIENCE

This book has spent twelve chapters dismantling a story. The story is that the systems running our water, our power, our pipelines, and our factories are besieged by an array of brilliant, patient, well-funded adversaries, and that the appropriate response is to spend whatever it takes to find and outmaneuver them, exploit by exploit, threat actor by threat actor. The fear machine sells that story because the story sells product, budgets, and headlines. What the fear machine cannot sell, because there is no recurring revenue in it, is the unglamorous truth this chapter is built on: most of what goes wrong in operational technology does not require an adversary at all, and the systems that survive are not the ones with the longest list of defeated attackers but the ones engineered to fail gracefully and recover fast no matter what knocks them over.

That is the argument of this final substantive chapter, and it is the book's positive thesis rather than its critique. If Chapter 8 was the diagnosis, the unmasking of the dull, accumulated, internal weaknesses that do the overwhelming share of real damage, this chapter is the prescription. And the prescription is not "patch faster" or "buy a better intrusion-detection box." It is a change of paradigm. The prevailing model in cybersecurity, and nowhere more so than in operational technology (OT), has for too long been *threat-centric*. It fixates on identifying and cataloging potential adversaries, the state-sponsored actor, the sophisticated criminal syndicate, the ideologically motivated hacktivist, and then constructing defenses tailored to counter their perceived capabilities and tactics. The narrative revolves around advanced persistent threats (APTs), zero-day exploits, and an arms race between attackers and defenders. Understanding adversaries is not worthless. But organizing an entire security posture around them produces systems that are brittle in exactly the ways that matter, and it does so at ruinous expense.

The alternative is *consequence-centric*. Instead of asking "who might attack us and how," it asks "what must keep working, what happens when it doesn't, and how fast can we bring it back." The threat-centric question has an infinite number of answers, because there is no end to the adversaries one can imagine and no end to the tactics they might invent. The consequence-centric question has a finite, tractable answer, because the number of critical functions in a water plant or a substation is finite, the ways they can degrade are enumerable, and the recovery paths can be designed, rehearsed, and measured. This reframing is the reason this chapter exists. It ties directly back to the spine of this book: the fear machine sells threat-chasing; engineered continuity is the evidence-based alternative.

Consider why the threat-centric approach fails on its own terms. It is fundamentally reactive. It builds walls against enemies who constantly evolve. Once a threat vector is mitigated, attackers shift to a new vulnerability or method, and the defender begins again. This is the "whack-a-mole" dynamic, and it produces a security posture that is expensive, exhausting, and fragile. Now picture an industrial control system painstakingly hardened against a specific family of malware. If that system is knocked offline not by malware but by an unexpected power surge during a storm, or by a cascading fault from an unrelated component, its sophisticated malware defenses are simply beside the point. The plant is down. It was not compromised by the anticipated threat; it was felled by an unforeseen systemic failure that the threat-centric posture never accounted for, because that posture was looking outward for an enemy when the failure came from within.

This is not a hypothetical. It is the documented pattern of the field. Chapter 8 made the case in detail, and I will not re-litigate it here: the weaknesses that actually get exploited in OT are overwhelmingly mundane, the default password never changed, the device never meant to face the internet that nonetheless does, the single VPN credential with no second factor behind it that gated a major regional fuel supply. The chief executive of the leading industrial-security vendor has stated flatly, in 2023, that there have been zero known ICS vulnerabilities ever leveraged in any ICS cyberattack.[1] Strip the threat narrative down to what the independent record will actually support, and the set of cyberattacks with confirmed cyber-physical effects remains small and category-dependent: Stuxnet caused physical equipment destruction; the 2015 and 2016 Ukraine grid attacks caused bounded service disruptions; TRITON targeted a safety-instrumented system and was discovered after a fail-safe trip; and disputed or bounding cases such as the German steel-mill report, FrostyGoop, and PIPEDREAM/Incontroller show why capability, service disruption, safety-system targeting, and verified equipment damage must be kept distinct.[2] The most rigorous count of nation-state OT physical attacks worldwide, meanwhile, runs to single digits and is falling year over year.[3] The point of recalling those figures here is not to deflate the threat again; Chapter 8 did that. The point is that a posture obsessed with the rare and spectacular is, by construction, a posture inattentive to the common and the survivable, and it is the common and the survivable that determine whether the lights stay on.

The threat-centric focus produces a narrow and incomplete assessment of risk. Worse, it can manufacture a false sense of safety by addressing the feared threat while leaving more probable failure modes wide open. A power utility may pour money into shielding its SCADA systems from nation-state espionage while neglecting the accidental misconfiguration, the aging hardware, or the insider action that can halt operations without any hacking skill whatsoever. The "threat," in the cases that actually recur, is rarely a shadowy foreign intelligence service. It is human error, equipment degradation, a botched software update, fatigue at the end of a long shift. These causes are undramatic, which is precisely why the fear machine ignores them and precisely why a consequence-centric posture must center them: their consequences are identical to, and far more frequent than, the consequences of the marquee cyberattack.

Chasing advanced threats is also disproportionately expensive relative to the probability of the attack. Sophisticated actors exist, but the typical OT system is far likelier to be disrupted by opportunistic ransomware spilling over from the business network, or by an internal failure, than by a bespoke cyber-physical payload. A posture organized around graceful degradation and rapid recovery allocates scarce resources where the expected harm actually lives. It invests in robust design, in redundancy, in fast restoration, in capabilities that pay off regardless of whether the disruption's source is an attacker, an operator, a thunderstorm, or a failed bearing. Money spent on fault tolerance is money that works against every cause of downtime at once. Money spent chasing a single adversary's tactics works against that adversary until the morning they change them.

There is a cultural cost to the threat-centric model as well. When the organizing principle is the external enemy, organizations hoard information and discourage candid conversation about their own weaknesses. Disclosing a vulnerability feels like painting a target on one's back. This instinct corrodes the very thing that makes an ecosystem stronger over time, the open exchange of what went wrong and why. If one operator discovers a novel weakness, the threat-centric reflex is to patch it quietly and tell no one, lest the disclosure invite attention. A continuity-oriented posture inverts the incentive: it treats every incident, near miss, and failure as data to be shared, because the goal is not to look invulnerable to adversaries but to make the whole sector harder to knock down. The contrast will recur throughout this chapter, because culture, not technology, is where most continuity programs ultimately succeed or fail.

The threat-centric model further tends to neglect foundational security in favor of exotic countermeasures. Patches slip because the maintenance window is inconvenient. Multifactor authentication is waved off as too cumbersome. Network segmentation is deferred as too complex. Each of these basic-hygiene failures widens the attack surface, and a posture preoccupied with the advanced threat is perfectly capable of justifying the neglect of the basics, on the theory that the real danger is elsewhere. The consequence-centric view reverses the priority. It treats the boring controls as load-bearing, because they reduce the surface across which *any* disruption, malicious or accidental, can propagate.

The march of technology only sharpens the contrast. The adoption of Industrial Internet of Things (IIoT) devices, cloud connectivity, and deepening IT/OT convergence is expanding the attack surface substantially. (The deeper structural reasons OT and IT pull against each other are taken up in Chapter 8 and the governance discussion later in this chapter; the relevant fact here is simply that the surface is growing faster than anyone can guard it.) An expanding, heterogeneous, perpetually changing surface makes the threat-centric ambition, defend against every attack, not merely difficult but incoherent. It demands the opposite move: away from preventing every incursion and toward building systems that can absorb a shock, preserve their essential functions, and come back quickly.

Take a water treatment plant. A threat-centric program fixates on preventing a sophisticated cyberattack that poisons the supply, a real concern that nonetheless crowds out more immediate and more probable risks. What about the failure of a critical pump worn out by age and deferred maintenance? What about an accidental software update that corrupts the control logic governing chlorine dosing? What about a tired operator's simple mistake that produces a discharge violation? Each of these can carry severe public-health consequences with no attacker anywhere in the picture. A continuity-oriented strategy answers all of them with the same architecture: redundancy in critical components so a single pump failure does not stop the plant; robust monitoring and alarms tuned to deviations from normal operating parameters so that a drifting chlorine level is caught long before it becomes a hazard; comprehensive operator training; and rehearsed response plans that cover the whole spectrum of failure, from a minor glitch to a major outage. Notice that none of these measures cares what caused the deviation. That indifference to cause is the entire point.

Even the definition of "threat" becomes a distraction inside the threat-centric frame. Governments and defense agencies categorize threats according to geopolitical rivalry and intelligence assessment, which steers policy toward defending against nation-state actors and away from the vulnerabilities that ransomware crews and insiders actually exploit. Nation-state capability is genuine. But the economic and societal disruption produced by ordinary ransomware sweeping across businesses and municipalities is, in aggregate, more immediate and more pervasive than the rare state-run operation. A consequence-centric posture insists that the *impact* of a disruption is what matters, and that mitigation should be prioritized by the potential for harm rather than by the perceived sophistication or exotic origin of whoever, or whatever, caused it.

The legacy character of OT infrastructure compounds the problem. Systems designed decades ago, with security an afterthought if a thought at all, are now being retrofitted with connectivity and modern interfaces they were never meant to carry. Building a threat-specific defense for so heterogeneous and fast-moving an environment is a losing campaign, like raising a dam against a coastline that will not hold still. A posture built on the inherent properties of the system, its ability to withstand stress, its redundancy, its fail-safe behavior, its capacity for rapid recovery, is durable in a way that any specific defensive technology is not. Those properties do not go obsolete the way a signature database or a threat-intelligence feed does. They are good against the next attack and against the next storm and against the next maintenance error, indefinitely.

The difficulty of predicting future threats is the final indictment of the threat-centric model. Novel adversary tactics, techniques, and procedures (TTPs) emerge with little or no warning. A system built for graceful degradation, with multiple defensive layers, isolated segments, robust recovery procedures, and worked-out contingencies, is far better positioned to ride out an exploit nobody anticipated than a system that depends on having seen the threat before. When a wholly new class of attack appears, the well-engineered system weathers it because it was never relying on foreknowledge of the attacker in the first place; the purely threat-centric system, by contrast, finds itself exposed precisely where its intelligence ran out.

A purely threat-centric approach also handles the human element badly. Insider risk, when it is considered at all, tends to be cast as a malicious actor, an internal version of the external enemy. Yet the reality is that most OT disruptions trace to accidental actions, gaps in training, fatigue, or simple error. A threat-centric program concentrates on detecting unauthorized access; it is far less likely to invest in the comprehensive training, the clear operating procedures, and the just culture, the environment in which people report near misses and minor errors without fear of reprisal, that actually prevent human-induced failures. Building robustness means strengthening exactly those human factors, equipping and supporting operators and putting processes in place to prevent and contain the mistakes that are, in the aggregate, every bit as disruptive as any cyberattack and far more common.

So the threat-centric approach yields a fragmented and less effective strategy. It spends vast resources mitigating known risks while systemic weaknesses go unexamined. It breeds a false confidence grounded in the mere absence of a specific, high-profile threat, leaving systems exposed to the unglamorous disruptions that do the actual damage. The constant pursuit of the latest threat distracts from foundational hygiene and from the operational integrity on which everything else depends. A consequence-centric, continuity-oriented posture offers a more pragmatic and more sustainable path to securing the OT systems that hold up modern life. The rest of this chapter builds that posture out, first as a design framework, then through the human factors that animate it, then through the foundational controls that implement it, then through the metrics that prove it is working, and finally through the governance reforms required to make it the default rather than the exception.

A Framework for Systemic Resilience

It helps to fix terms. Throughout this chapter I will speak of a system's *robustness* (its ability to take a hit without breaking), its *graceful degradation* (its ability to shed nonessential function while preserving the critical core), its *fault tolerance* and *redundancy* (its ability to keep running when a component dies), its *recoverability* (the speed and completeness with which it returns to normal), and its *continuity* (the uninterrupted delivery of essential service across all of the above). These are not synonyms for one big word. They are distinct engineering properties, each measurable, each purchasable with specific design decisions. When the whole bundle holds together under stress, we may fairly call the result *resilience*, the capacity of a system to anticipate, withstand, recover from, and adapt to disruption while maintaining its critical functions. I will reserve that term for the bundle rather than spend it on every individual control, because a word applied to everything ends up describing nothing.

The defining feature of this framework is that it is organized around consequences, not causes. A threat-centric design begins with a catalog of adversaries and works backward to controls. A continuity-oriented design begins with the critical functions that must not fail, asks how they can degrade, and works forward to the redundancy, failsafes, procedures, training, and recovery paths that keep those functions alive under stress, whatever the stress turns out to be. The cause might be a remote intruder, a lightning strike, a failed power supply, a corrupted firmware push, or an exhausted operator. The design does not need to know in advance. That is its great economy: one architecture, hardened once, that pays off against the entire population of disruptions rather than against a single imagined attacker. What follows are the core principles of that architecture, spanning technology, process, and people.

Redundancy Redundancy is the deliberate provisioning of duplicate or backup components, processes, or facilities so that the failure of one element does not become the failure of the function it served. It is the most literal embodiment of the consequence-centric stance, because it makes no assumption about *why* the primary failed. A redundant processor does not ask whether its partner was crashed by malware or by a cosmic ray; it simply takes over. In OT settings this principle shows up at every layer. Critical control systems may run redundant processors or communication links, so that if the primary processor faults, or is compromised, the secondary assumes control seamlessly and downtime is minimized or eliminated. Critical network infrastructure such as switches and routers can be deployed in high-availability pairs, a standby unit activating the instant the active unit fails.

Redundancy extends well beyond processors. It applies to software, to data, and to power. Maintaining multiple, independent backups of critical configurations, operational data, and control logic is foundational, and those backups are worthless unless they are stored securely and, crucially, tested on a schedule; an untested backup is a hypothesis, not a safeguard. The principle reaches power as well, where uninterruptible power supplies (UPS) and backup generators form a layered defense against the outage, which, it bears repeating, is far more often caused by the grid or the weather than by any attacker. For larger operations, geographic redundancy, duplicate control centers in separate locations, guards against a localized disaster taking everything at once. Across all of these forms, the property that matters is the same: the transition from primary to backup must be automatic and fast, because a failover that depends on a human noticing and reacting in the middle of the night is a failover that will eventually arrive too late.

Fail-Safe Design Where redundancy keeps a function running, fail-safe design governs what happens when running is no longer possible. The principle holds that a system, when it does fail, should default to a safe state rather than an arbitrary one. A fail-safe system does not merely stop; it actively moves toward the condition that prevents harm to people, equipment, and the environment. This is consequence-centric thinking in its purest form, because a safe default protects against the entire universe of failure causes simultaneously. In industrial automation it usually takes the shape of "fail-open" or "fail-closed" behavior chosen according to the safety requirement. A valve metering a hazardous chemical may be engineered to fail-closed, shutting off flow automatically on loss of power or control signal and so preventing a spill, whether that loss was caused by a cyber intrusion, a tripped breaker, or a severed cable. A safety shower, conversely, may be designed to fail-open, guaranteeing it activates if its control mechanism malfunctions.

Beyond the individual component, fail-safe principles operate at the level of the whole system. Control loops can be designed with inherent safety parameters; watchdog timers can monitor critical processes and trigger an orderly, safe shutdown when anomalies appear. Embracing fail-safe design also means studying how a system can fail before it does, through structured techniques such as Failure Mode and Effects Analysis (FMEA). The value of FMEA in a consequence-centric program is that it enumerates failure modes and their effects without privileging any particular cause — adversarial or otherwise: it enumerates the ways a function can be lost and the effects of each loss, without first demanding a threat actor to blame. That discipline surfaces risk early, in the design phase, where it is cheapest to address, and it makes the system's behavior during an incident predictable and safe rather than chaotic. Predictability under stress is itself a form of protection, because it is the cascading, unforeseen secondary failure, the one the designers never modeled, that turns a contained problem into a catastrophe.

Robust Operational Procedures Technology supplies the capacity for continuity; procedures convert that capacity into reliable practice. Robust operational procedures are the documented, standardized methods for running and maintaining OT systems, and they are where many continuity programs quietly live or die. They span everything from daily startup and shutdown sequences to routine maintenance, emergency response, and change management. To support continuity rather than merely to satisfy an auditor, these procedures must be comprehensive, clear, and genuinely accessible, leaving no room for ambiguity at the moment someone under pressure reaches for them. They must say explicitly how to handle deviations from normal conditions, how to respond to alarms, and when and how to escalate. And because the systems and the threats and the lessons all change, the procedures must be reviewed and updated on a cycle that keeps pace, incorporating what was learned from every incident and every near miss.

A chronic pitfall in OT environments is reliance on informal, undocumented knowledge held in the heads of a few experienced operators. That tribal knowledge is valuable, but it is fragile: it walks out the door when a veteran retires, and it varies from person to person in ways that defeat consistency. Formalizing it into robust procedures is itself an act of continuity engineering, a hedge against the loss of a single point of human failure. Rigorous change management belongs here too. Any modification to the OT environment, a software update, a hardware replacement, a configuration change, must be planned, tested, and authorized, with a risk assessment that asks how the change could affect the system's ability to withstand and recover from stress, and with a rollback plan ready in case the change itself becomes the disruption. More than one outage has been self-inflicted by a well-intentioned update pushed without a way back; the rollback plan is the redundancy principle applied to change itself.

Comprehensive Training Technology and procedures are inert without people equipped to use them. Training for OT environments must reach past basic operational tasks to a working understanding of the systems, their failure modes, and the reasoning behind every continuity measure. It must cover normal operation, troubleshooting, emergency response, cybersecurity hygiene, and the correct use of fail-safe mechanisms, not as a checklist of rules but as an understanding of why the rules exist, so that an operator can act sensibly when the situation in front of them does not match any rule they were given. The deepest aim of continuity training is to cultivate a culture of continuous learning and vigilance, an environment in which operators are encouraged to report anomalies, near misses, and potential safety or security concerns without fear of reprisal, because each such report is early warning that a function may be drifting toward failure.

Drills and simulations are indispensable here, and the consequence-centric framing shapes what they should rehearse. They should not drill only the cyberattack. They should mimic the full realistic spectrum, the cyber intrusion, yes, but equally the equipment failure, the environmental challenge, the bad data, the loss of a communication link, so that personnel practice their roles against the kinds of disruption that actually recur and not merely against the kind that makes headlines. Training should be tailored to role, from frontline operator to senior management, so that everyone understands what they are responsible for when the system degrades. And because the systems and their failure modes keep evolving, training cannot be a onetime event; it demands refreshers and specialized instruction as new risks, of every kind, emerge.

Rapid Recovery Capabilities If redundancy and fail-safes determine how well a system withstands a shock, recovery determines how fast it comes back, and recoverability is the dimension where the consequence-centric posture most clearly outperforms the threat-centric one. The threat-centric mindset treats recovery as the thing you do once prevention has failed, an admission of defeat. The continuity-oriented mindset treats recovery as a first-class capability to be designed and rehearsed in advance, on the understanding that some disruption, from some cause, is inevitable, so the question that matters is not whether you will ever be knocked down but how quickly you can stand back up. Rapid recovery requires pre-planning: clear, executable strategies; incident response plans that lay out the steps, the roles and responsibilities of the response team, the communication protocols, and the escalation path. And these plans must be tested and refined on a schedule, because an untested recovery plan is of limited value, a comforting document that has never met reality.

Testing backups and restoration is the heart of recoverability. It is not enough to possess backups; one must be confident they can be restored within an acceptable window and that the restored data and configurations are accurate and functional. This is where recovery time objectives (RTOs) and recovery point objectives (RPOs) earn their keep, defined to align with what the business and the public actually require. For the most critical OT functions, RTOs can be tight — sometimes measured in minutes — which forces a high degree of automation, automated system deployment, infrastructure as code, pre-configured recovery environments, because a recovery process that depends on heroics and improvisation at three in the morning is a process that will miss its objective. Equally vital is the ability to isolate a compromised or malfunctioning segment so that the rest of the plant or network keeps running while the affected portion is restored. That capacity to contain and recover a part without sacrificing the whole is graceful degradation made operational, and it is the difference between an incident and a disaster.

Strong Physical and Cyber Defenses A continuity-oriented posture does not abandon defense; it puts defense in its proper place. Robust physical and cyber defenses still matter, because preventing or slowing a disruption is cheaper than absorbing and recovering from it, and these defenses remain nonnegotiable. The shift is in the reasoning behind them. We do not deploy a firewall because we have personified a particular hacker and mean to thwart him. We deploy it because it shrinks the surface across which any disruption, intrusion, misconfiguration, or malware spillover, can spread, and because a smaller surface is a more survivable one regardless of what comes at it. In OT environments physical security is as critical as its cyber counterpart, controlling access to facilities, securing control rooms, protecting hardware, and implementing surveillance and intrusion detection.

Cyber defenses must complement those physical controls, and the convergence of IT and OT networks, which blurs the line between the physical and the digital, makes a holistic approach essential rather than optional. Defenses in OT must be tailored to the peculiarities of the environment, legacy systems, real-time constraints, specialized protocols. That tailoring can include network segmentation to isolate critical control networks from less trusted enterprise networks; robust authentication and access control; intrusion detection and prevention systems (IDPS) sensitive to OT protocols; and regular vulnerability assessment and penetration testing. But there is a discipline specific to OT that a threat-centric program routinely forgets: the defense must not itself become a source of failure. A security control that impedes operational function, or that introduces a new single point of failure, has traded one disruption for another. The aim is a layered architecture that is robust by design, in which the compromise of one layer does not cascade into compromise of the whole, with controls at the network, host, and application levels and particular care for remote access points and third-party connections, which are disproportionately the route by which trouble, of every kind, gets in. Integrating security monitoring and logging across both the physical and cyber domains provides the comprehensive view of system health that lets an operator catch a deviation, whatever its origin, before it becomes an incident.

These principles, redundancy, fail-safe design, robust procedures, comprehensive training, rapid recovery, and properly scoped defenses, form the bedrock of a continuity-oriented OT infrastructure. Building on them, an organization moves from a reactive crouch to a system capable of withstanding the operational challenges it will actually face. Yet technology and procedures, however sound, remain fragile if the human element is neglected. Managing human factors is not a supplement to this framework; it is a pivotal part of it, reaching up into organizational culture and leadership, and it is to those factors that we now turn.

Well-Trained and Skilled Personnel

Automation can now manage a large share of routine operation. But the nuances of an industrial control system, the subtle precursor to a failure, the judgment call under pressure, lie largely beyond the reach of purely automated detection, and this is exactly where the consequence-centric posture leans hardest on people. A system designed to degrade gracefully and recover fast still needs a human who can recognize that it is degrading, often from signals that fall outside any predefined parameter, and who understands the system well enough to choose the response that preserves the critical function. It is the experienced operator, the diligent maintenance technician, the insightful supervisor, who interprets the ambiguous sensor reading and makes the call the automation cannot. This calls for investment in training that is capability-enhancing rather than merely compliance-driven, training that fosters a deep, intuitive grasp of system behavior, of failure modes, and of the interdependencies that turn a small fault into a cascade. That means hands-on simulation, scenario-based exercises that mimic realistic disruptions of every kind, and continuous development that keeps pace with evolving technology. The goal is a workforce that is not merely competent but adaptive, that understands the *why* behind the procedure and can improvise effectively when the unexpected arrives, because the unexpected, by definition, is what no procedure anticipated.

This depth of knowledge is what lets people catch problems early and decide well in a crisis, keeping a minor deviation from escalating into a major failure. A grid operator must understand the cascading effects of switching a circuit during a storm, not merely how to throw the switch. A water plant operator must understand the chemistry well enough to adjust dosing in response to a contamination event, not merely follow a rote step. The recurring storm and contamination examples are deliberate: the human capability that matters is the one that holds up against the everyday physical disruption, not just the exotic intruder.

Effective Leadership During Crises

When a disruption arrives, leadership's clarity, decisiveness, and communication can be the difference between a controlled recovery and a catastrophic one. Leaders set the strategic direction, allocate resources, make the hard tradeoffs, and create the conditions in which accurate information moves quickly to the people who need it. This demands competencies beyond routine management: the ability to stay calm under pressure, to analyze incomplete information fast, to delegate well, and to inspire confidence. It also demands a working familiarity with the organization's continuity plans and the capacity to adapt them on the fly as the situation shifts, which is why a leader who has actually participated in continuity planning and exercises will be far better equipped to steer an organization through a real event than one encountering the plan for the first time when it matters most.

Crisis leadership also depends on robust communication. Leaders must push clear, actionable information to their teams, their stakeholders, and the public while pulling critical updates from the front lines. In OT this might be a plant manager ordering a controlled shutdown, the fail-safe decision deliberately taken, or a utility executive coordinating with government agencies during a power outage whose cause may not even be known yet. The authority to make such calls, paired with the workforce's trust, is itself a continuity asset. Without strong leadership, organizations descend into the chaos that prolongs recovery and breeds the secondary failures that turn a single problem into many.

A Strong Organizational Culture

The disposition to withstand and recover from disruption is not a feature bolted onto an organization; it must be woven into its values and its daily operation. That means an environment in which safety and security are understood not as burdens on productivity but as intrinsic enablers of sustained operation, an environment that actively encourages the open reporting of near misses and potential hazards. This is the cultural inversion the threat-centric model cannot perform. A culture organized around external enemies hoards information and punishes the disclosure of weakness; a culture organized around continuity treats every disclosure as a gift.

At the center of it sits the *just culture*, in which individuals can report errors or flag vulnerabilities without fear of retribution. This is the single most important determinant of whether weaknesses get fixed before they fail rather than after. When people feel safe to report, the organization gains the insight to update procedures, refine training, and shore up defenses *before* a major incident, rather than reconstructing what went wrong in the aftermath. A technician who notices a recurring alarm on a pump, an alarm long dismissed as a nuisance, may, in a just culture, feel empowered to investigate and uncover a developing mechanical fault, heading off a costly and disruptive failure. In a culture where reporting is discouraged or punished, that same signal is suppressed, and the small fault is permitted to grow into the large one. Note that the saved failure in that example is mechanical, not malicious; the just culture pays off against the everyday causes of downtime, which is exactly why it belongs at the center of a consequence-centric program.

This cultural shift also embeds cybersecurity awareness at every level of operation. In OT environments the historical divide between IT and OT security can produce a siloed posture in which OT personnel do not fully appreciate the cyber risks they run. A continuity-oriented culture ensures that everyone understands their part in keeping critical systems running, from locking down a control panel to adhering to strict protocols for remote access and software updates. That pervasive awareness transforms security from a specialized IT function into a shared responsibility, which is precisely what a consequence-centric posture requires, since the consequences land on everyone regardless of which silo the cause originated in. Leadership shapes this culture, not through pronouncements but through consistent action, through how it allocates resources and makes decisions. When leadership visibly prioritizes safety and continuity investment even under budget pressure, it sends a message that reinforces the culture far more powerfully than any policy memo.

Personnel Selection and Development

Traits suited for critical infrastructure roles, such as attention to detail, methodical thinking, and the ability to remain calm under pressure, should be key considerations during recruitment. Once onboard, continuous investment in their development is paramount. This includes providing opportunities for cross-training, allowing personnel to gain experience in different aspects of the system. For example, a control room operator might benefit from spending time with maintenance crews to understand the physical realities of the equipment they monitor. This cross-pollination of knowledge builds a more holistic understanding of the system's vulnerabilities and operational dynamics.

Organizations must also actively manage knowledge transfer, because the departure of an experienced operator is itself a disruption, a loss of tacit knowledge that no adversary inflicted but that degrades the system's capacity to withstand and recover all the same. Mentoring programs, knowledge-sharing platforms, and serious succession planning are the redundancy principle applied to people: they ensure that critical expertise does not reside in a single irreplaceable head. Developing future leaders matters for the same reason. Identifying individuals with leadership potential and giving them specialized training in crisis management and strategic thinking ensures the organization will have capable stewards when the next disruption, whatever its cause, arrives. This proactive approach to talent is a strategic investment in the organization's ability to keep operating safely under stress, and it pays off against turnover, retirement, and the merely human limits of any one expert, none of which a threat-centric program ever budgets for.

The effectiveness of every human factor described so far is, in turn, profoundly shaped by the design of the interfaces through which people actually perceive and act on the system.

Human-Machine Interface (HMI) Design

The design of human-machine interfaces (HMIs) and control systems directly shapes performance, decision-making, and error rates, which makes the HMI a continuity component in its own right rather than a cosmetic concern. A poorly designed HMI manufactures its own disruptions: it confuses the operator, invites the misreading of data, and piles on cognitive load until a manageable situation becomes an operational failure, no attacker required. A well-designed HMI, by contrast, presents information clearly, intuitively, and with the right prioritization. Critical alarms are distinct and unambiguous; high-risk control actions are intuitive yet demand deliberate confirmation; color, symbology, and layout obey established ergonomic principles and stay consistent across the system. In a complex power-distribution control room, a good interface uses color coding and graphical representation to convey at a glance the status of substations and transmission lines, letting operators spot a fault or a developing problem quickly, precisely the early recognition on which graceful degradation depends. An HMI that buries the operator under undifferentiated data, uses confusing icons, or demands elaborate action sequences for simple tasks degrades performance and, in doing so, becomes a latent cause of the very incidents continuity engineering is meant to prevent. This is acutely relevant in OT, where legacy interfaces are common and hard to update.

Organizations should therefore prioritize modernizing HMIs, designing them around the limits of human cognition through user-centered design, extensive testing with real operators, and a commitment to iterative improvement. The aim is an interface that actively supports informed decision-making under pressure. The connection among technology, procedure, and people is the thread through all of it: technology provides the tools, procedures supply the roadmap, but it is the human element that navigates the journey, and in high-stakes OT environments a failure to attend to human factors will undermine even the most robust technical defenses. Organizations must treat their workforce as active participants in the continuity architecture, not as a residual risk to be controlled. Investments in human capability, in culture, and in interface design are not incidental expenses; they are fundamental to a system's ability to withstand and recover from disruption. The human element, properly nurtured, becomes the most potent defense and the most agile recovery mechanism an organization has, working against every cause of failure rather than against a single imagined enemy.

A continuity-oriented posture rests, finally, on a bedrock of foundational security practice. Advanced countermeasures have their place, but a preoccupation with them, the signature failing of the threat-centric model, diverts resources from the unglamorous disciplines that actually prevent the majority of disruptions. Genuine robustness demands a sustained commitment to baseline controls whose collective effect is far greater than the sum of their parts, and the argument for each of these controls, reframed here, is not "do this because hackers" but "do this because failure is inevitable from many causes and these controls shrink the surface across which any of them can spread." The first of those pillars is diligent patch management.

Diligent Patch Management

Software contains defects; this is a permanent condition, not a passing threat, which is why routine patching is a continuity practice rather than merely an anti-hacker measure. Weaknesses surface through vendor disclosure, independent research, or active exploitation, and any of them can serve as an entry point, but the deeper reason to patch is that an unmaintained system accumulates fragility from every direction at once. In OT the stakes are amplified by the direct coupling to physical process: an unpatched flaw in a PLC (Programmable Logic Controller) or a SCADA (Supervisory Control and Data Acquisition) system could let a fault, malicious or otherwise, propagate into manipulated sensor readings, altered control parameters, or a halted process. A systematic approach to patching is therefore essential, and it is worth being honest, as Chapter 8 was, that the threat narrative around ICS vulnerabilities is wildly overstated; the case for patching here rests not on the fantasy of the air-gap-leaping zero-day but on the cumulative reduction of surface and fragility.

A systematic approach is more than finding patches. It requires robust processes for testing them in non-production environments so the patch itself does not become the disruption, a real risk in OT, where a careless update can destabilize a running process more reliably than any attacker. For critical infrastructure this testing phase is paramount and demands close collaboration among IT, OT, and vendor teams. Deployment must be orchestrated carefully, often through phased rollouts and predefined rollback plans inside planned maintenance windows, the rollback being, once again, the redundancy principle applied to change. The frequency of patching should be governed by risk assessment, prioritizing the vulnerabilities most likely to matter in the specific environment rather than chasing every disclosed CVE, which is itself a small act of consequence-centric discipline. Neglect timely patching and the surface widens silently; the cycle of discovery and remediation is a perpetual necessity, and organizations that ignore it pay more later, in downtime from causes they never bothered to name.

Network Segmentation

Modern OT networks are densely interconnected, and that interconnection, convenient for operations, is also a highway down which any disruption can travel fast, a worm, a misconfiguration's blast radius, ransomware spilling from the business side. Network segmentation divides the network, logically or physically, into smaller, isolated zones, each with its own security policy and access controls. The spread of IIoT and the deepening convergence of IT and OT have made this harder, blurring traditional boundaries and opening new surface; the structural tension between IT and OT priorities behind that convergence is examined in Chapter 8. Segmentation is, at its core, the containment principle of graceful degradation expressed in network terms: when one zone is compromised or fails, the trouble is prevented from cascading into the critical areas, so the system loses a part rather than the whole. This is the logic of "defense in depth," multiple layers arranged so that no single breach is catastrophic. A power-generation facility's control network is strictly segmented from its business network; within the OT environment, safety instrumented systems (SIS) are, per IEC 61511, kept independent of general process-control networks to the extent practicable, precisely so that a problem in one cannot disable the safety function that exists to bring the plant to a safe state.

Effective segmentation requires deliberate network architecture, the deployment of firewalls at the boundaries, and strict enforcement of communication policies that permit only necessary traffic. The Purdue Model offers a valuable framework here, defining distinct zones for different control levels; each zone carries defined security requirements, and traffic crossing a boundary is inspected. The payoff is twofold and, again, indifferent to cause: segmentation limits an attacker's lateral movement and, in the same stroke, contains the blast radius of an accidental fault and improves network performance besides. One architectural decision, many disruptions contained.

Secure Configuration Management

Secure configuration management is the next foundational layer. Every device and system in an OT environment, from servers and workstations to PLCs and network switches, ships with a default configuration, and those defaults are tuned for ease of use and broad compatibility rather than for survivability, frequently enabling unnecessary services, protocols, and access mechanisms. Secure configuration management is the meticulous hardening of those systems: disabling nonessential services, removing default credentials, enforcing strong password policy, closing unnecessary ports. The connection to Chapter 8's diagnosis is direct and worth stating plainly, because the default credential left in place is not a hypothetical the fear machine conjured; it is, by available incident reporting, a recurring way real OT systems are compromised, and hardening is the cheapest, most cause-agnostic defense against it. For OT this discipline is especially critical, since many legacy devices cannot easily be reconfigured or patched and so must be configured correctly from the outset and kept that way.

Using configuration management tools to establish a known-good baseline for each device is essential, as are regular audits against those baselines to detect unauthorized drift from the secure state, drift being itself a quiet form of degradation that, left unnoticed, erodes robustness over time. A robot controller's diagnostic ports, for instance, might be disabled whenever they are not in use. In environments where systems are deployed for decades, maintaining secure configuration is not a onetime task but an ongoing one; an unattended system with weak credentials is an open invitation, and not only to a sophisticated attacker but to the opportunist and the accident alike. Proactive hardening shrinks the surface and makes a foothold, of any kind, harder to establish.

Strong Access Controls

The principle of least privilege holds that users, processes, and systems should be granted only the minimum access necessary to do their jobs, and its value in a consequence-centric program is that it limits the damage any single failure can do regardless of whether that failure is a malicious insider, a compromised account, or an honest mistake by someone who simply had more reach than the task required. In OT this means ensuring that only authorized personnel can touch a given control system, dataset, or function, through strong authentication, such as multi-factor authentication (MFA) where it is feasible, and robust authorization policy. Role-based access control (RBAC) is the common and effective strategy, assigning permissions to roles rather than to individuals, which simplifies management and enforces consistency. OT access controls must be granular: an operator might view system status and acknowledge alarms but be unable to change critical setpoints or initiate a shutdown; maintenance personnel might receive temporary, time-bound, logged elevation for a specific task. Centralized identity and access management (IAM) can streamline all of this. The reframing matters, the legacy VPN account without multifactor authentication that Mandiant identified as the initial access vector in the 2021 Colonial Pipeline ransomware incident, which prompted a precautionary shutdown of a major U.S. fuel pipeline, discussed in Chapter 8, was not defeated by brilliance; it was a missing control, and the disruption that followed was magnified by the absence of segmentation behind it.

Privileged accounts, those with broad rights, demand special attention and must be strictly controlled, monitored, and audited. The insider risk, whether malicious or accidental, is a serious concern, and strong access controls are a primary defense against both faces of it; the consequence-centric posture refuses to treat the insider purely as an enemy when the more frequent reality is the well-meaning person who erred. Physical security for control rooms and equipment must be aligned with these digital measures, since an access control that stops at the network and ignores the door is a badly incomplete one.

Regular Vulnerability Assessments

Regular vulnerability assessment and penetration testing provide the feedback loop that tells an organization whether its foundational controls are actually holding. Patching, segmentation, and access control are preventive; assessment and testing are how one learns whether they work before a real disruption delivers the verdict. Vulnerability assessment systematically scans systems and networks for known flaws, combining automated tooling with manual analysis to surface the subtler weaknesses.

Penetration testing goes further, simulating real attacks to actively exploit identified weaknesses, and in OT it must be conducted with extreme caution, because the test itself can become the disruption it was meant to anticipate. The potential for unintended operational impact means such tests are often confined to scheduled maintenance windows, coordinated closely with operations, and sometimes executed with specialized, non-intrusive techniques. The goal is not to cause damage but to learn how an attacker might gain unauthorized access or control, and the same exercise frequently exposes brittleness that has nothing to do with attackers at all, the fragile dependency, the single point of failure, the recovery path that does not work as documented. The results guide the prioritization of remediation and reveal controls that are insufficient or misconfigured. It is a cyclical discipline: assessment reveals weakness, remediation strengthens, the next assessment validates the improvement. Without these regular checks an organization drifts into false confidence, unaware of the misconfiguration or the emerging exposure that could, one day, hand a routine event the leverage to become a devastating one.

Investing in these foundational practices is a strategic imperative for operational continuity. "They are the essential, largely unheralded disciplines that, by the available evidence, prevent most security incidents in OT," and they do so precisely because they are cause-agnostic: by resourcing routine patching, effective segmentation, secure configuration, robust access control, and diligent assessment, an organization shrinks its attack surface, contains the impact of whatever breaches or failures do occur, and establishes a posture that is far more cost-effective and reliable than perpetually chasing the latest sophisticated threat. Ultimately it is about building a foundation that can withstand the inevitable pressures, from every direction, and keep critical infrastructure running. Yet foundational practice, however sound, is not the end of the matter, because executing controls is not the same as knowing whether they have made you more survivable. For that, we need to measure the thing itself, not adherence to a checklist but the capacity to withstand and recover, which is the subject of the section that follows.

The capacity to withstand and recover from disruption can be measured, and measuring it well requires abandoning the question the threat-centric, compliance-driven world keeps asking, "Are we compliant?", in favor of the central question a consequence-centric posture cares about: "How well can we operate when things go wrong?" The distinction is not pedantic. Compliance metrics certify that you have installed the controls; continuity metrics tell you whether the system actually survives.

Consider the analogy of a building. A compliance metric checks whether the structure meets current code for earthquake resistance. That is worth knowing, but it tells you nothing about how much damage the building will sustain in a particular quake, how fast it can be repaired, or whether it can be occupied afterward. Continuity metrics aim instead to quantify performance under stress: how much the building sways, which structural elements fail, how long repairs take, the timeline to full reoccupation. Translated to OT, this means metrics that reflect the actual operational capacity of a facility or a sector under duress, not the box-checking of an audit. To get there we need quantifiable measures for the four dimensions named at the start of this chapter, the capacities to anticipate, withstand, respond, and recover, which together constitute what we are entitled to call resilience. Each dimension is treated in turn below, and the through-line is consistent: every metric asks about consequence and capacity, not about whether a specific adversary was thwarted.

Anticipate: Foreseeing Threats

This dimension concerns an organization's ability to foresee disruptive conditions and prepare for their impact, and the consequence-centric reframing widens its aperture deliberately, since the goal is to anticipate failure of every kind and not merely the next adversary. Metrics here should reach past the raw vulnerability-scan tally. **Threat Landscape Intelligence Maturity:** rather than simply counting known vulnerabilities, we measure the maturity of the organization's intelligence and foresight program. Useful measures include:

Proportion of known exploitable vulnerabilities actively tracked and prioritized: moving past the mere existence of a vulnerability to its immediate relevance and realistic exploitability within the specific environment, which is itself an exercise in consequence over abstraction, since most disclosed flaws, as Chapter 8 notes, are not reachable in practice.

Timeliness of intelligence integration into risk assessments: how quickly new information, about adversaries but also about failure modes, equipment recalls, and environmental hazards, is folded into the organization's understanding of its own exposure, measured for example by the average lag between a significant intelligence update and its incorporation into internal risk models.

Effectiveness of threat modeling and scenario planning: gauged by the number of plausible disruption scenarios, of every cause, that have been identified, mitigated, and actually tested through exercise or simulation, because a scenario never rehearsed is a scenario not truly anticipated.

Operational Readiness for Emerging Threats: how well the organization's systems and processes are built to adapt to new forms of disruption. Useful measures include:

Percentage of critical systems with adaptable security architectures: whether systems are designed with the modularity and flexibility to absorb new controls or configuration changes without a wholesale overhaul, systems that can, for instance, quickly implement a new network access policy or update a behavioral anomaly-detection model. Adaptability is the fourth dimension of resilience made concrete.

Rate of adoption of emerging security technologies relevant to OT: how quickly the organization pilots and integrates technologies proven against novel failure modes, such as advanced endpoint detection and response (EDR) for OT endpoints or anomaly detection tuned to industrial protocols.

Staff proficiency in emerging threat defense: measured through regular, realistic training and certification focused on current attack methodologies and on the defense and recovery techniques relevant to OT, the human-capability dimension carried into measurement.

Withstand: The Capacity to Absorb Impact

This dimension measures an organization's ability to endure a disruptive event with minimal degradation of critical function, to keep the essential core running under duress while shedding what can be safely shed. It is graceful degradation rendered as metric.

Systemic Robustness and Redundancy: moving past the reliability of individual components to the whole system's ability to maintain function.

Critical Function Uptime Ratios under Simulated Stress: testing whether a system holds its core functions during high load, partial failure, or degraded connectivity. A grid operator might measure the uptime of essential monitoring and control functions during a simulated severe-weather event that degrades communication links, an explicitly non-adversarial stress, chosen because the weather is a more frequent cause of that degradation than any attacker.

Effectiveness of Network Segmentation and Isolation: beyond the mere presence of segmentation, metrics should test how well it actually contains a problem. These could include:

Mean Time to Detect lateral movement across segments: how quickly a breach is detected if it crosses a perimeter segment and tries to move deeper, an indicator that doubles as a measure of how fast any spreading fault would be caught.

Proportion of critical assets that are logically or physically isolated from less secure zones: quantifying the real degree of isolation around the most sensitive components, which is to say the real extent of the containment on which graceful degradation depends.

Effectiveness of choke points: how well firewalls and intrusion prevention systems at segment boundaries block or detect unauthorized traffic during a simulated event.

Metrics can also gauge data integrity by tracking the rate of successful data-validation checks and the timeliness of anomaly detection for data manipulation, corruption being a failure mode whether it arrives by malice or by a faulty sensor. **Percentage of critical components with verified supply-chain security:** assessing the security posture of vendors and the integrity of the components they supply, moving past mere contractual compliance to actual provenance.

Availability of alternative suppliers or redundant supply chains for critical materials and components: quantifying the organization's ability to switch to backup sources if the primary supply is disrupted, the redundancy principle extended outward into the supply chain, where the disruption is at least as likely to come from a port closure or a bankruptcy as from sabotage.

Respond: Acting Swiftly

This dimension measures the speed and effectiveness of the actions taken to minimize an incident's dwell time and contain its spread, and the metrics below are pointedly indifferent to what triggered the incident; a fast containment of a spreading fault is valuable whether the fault is malware or a runaway process.

Incident Response Plan Effectiveness: the plan's real-world performance.

Mean Time to Detect (MTTD) Incidents: how quickly events are identified and acknowledged, a leading indicator of overall response capacity.

Mean Time to Triage Incidents: once detected, how quickly an incident's scope and impact are assessed.

Mean Time to Contain (MTTC) Incidents: how quickly an incident is stopped from spreading, which in OT can be measured by the time taken to isolate an affected segment or system without forcing an uncontrolled shutdown of critical processes, containment without collateral being the operational expression of graceful degradation.

Effectiveness of Automated Response Capabilities: for certain incident types, automated response can sharply reduce MTTC; a useful measure is the percentage of defined incident types for which an automated response is triggered and succeeds.

Communication and Coordination: effective response depends on seamless communication.

Timeliness and accuracy of communication to stakeholders: internal and external alike, measurable by adherence to communication timelines and the accuracy of what is shared.

Effectiveness of Cross-Functional Team Collaboration: during an incident, IT, OT, engineering, and management must act in concert, the dissolving of the silo discussed earlier made measurable. Metrics could include survey-based assessments of collaboration effectiveness after an incident, or the efficiency of joint decision-making.

Crisis Management Maturity: the organization's ability to manage the whole situation, not just the technical slice.

Regularity and realism of crisis simulation exercises: the frequency and complexity of exercises that test the entire crisis framework rather than the technical response alone, and, in keeping with the consequence-centric stance, exercises that rehearse the equipment failure and the natural disaster as seriously as the cyberattack.

Post-crisis debriefing and lessons-learned integration: the systematic capture of insight from incidents and exercises and its feedback into policy, procedure, and training, the just culture operationalized as a metric.

Recover: The Return to Normalcy and Beyond

This dimension measures how quickly and how completely an organization restores operation, and how well it learns from a disruption to become more survivable next time, the adapt capacity folded into recovery. Recoverability, recall, is where the consequence-centric posture most decisively outperforms the threat-centric one, because it accepts that being knocked down is inevitable and concentrates on standing up fast.

Traditional metrics such as Recovery Time Objective (RTO) and Recovery Point Objective (RPO) must be applied more rigorously than the audit typically demands.

Actual Recovery Time vs. Defined RTO for Critical Functions: how quickly essential services were actually restored against what was planned; the deviations are where the next improvement lives.

Resilience of Restoration Processes: the robustness of the backup and recovery machinery itself. Useful measures include:

Either change the label to "Failure rate of restoring from backups" or change the description to "how often a restoration attempt succeeds." Recommended: "**Success rate of restoring from backups:** how often a restoration attempt succeeds, the question that exposes the untested backup before the crisis does.", the question that exposes the untested backup before the crisis does.

Time to restore full functionality of redundant systems: when a primary fails and a backup activates, how long until the backup is operating at full capacity, the failover principle held to account.

Business Continuity and Disaster Recovery (BC/DR) Effectiveness:

Percentage of critical business processes that meet their defined RTOs and RPOs post-disruption: a high-level view of overall recovery success across the organization.

Effectiveness of workarounds and temporary solutions: during recovery, temporary measures are common; metrics can assess how well these workarounds sustain essential operation and how quickly they are replaced by permanent fixes, the workaround being graceful degradation persisting into the recovery phase.

Continuous Improvement and Learning:

Rate of implementation of corrective actions post-incident/exercise: the organization's agility in actually making the improvements its lessons-learned process identifies, since insight not acted upon is not adaptation.

Reduction in recurrence of similar incidents: over time, effective learning should drive down the frequency of specific kinds of disruption, a direct measure of whether the system is becoming more survivable.

Adaptation of resilience strategies based on evolving conditions: a qualitative but crucial measure of whether the organization's continuity framework is dynamic and responsive to new threats *and* to new operational realities of every other kind, equipment, climate, supply, workforce.

Implementing these metrics requires a cultural shift toward performance, investment in data analysis, and collaboration across IT, OT, and business units so that the numbers drive action rather than decorate a report. To be worth collecting, the metrics must be actionable, enabling an organization to:

1. **Identify Weaknesses:** pinpoint where the capacity to withstand and recover is lacking, such as slow detection, inadequate redundancy for a particular function, or a recovery procedure that has never been properly tested.

2. **Prioritize Investments:** justify resource allocation by demonstrating the quantifiable benefit of a continuity initiative. A metric showing a high MTTC for a particular incident type makes a strong, evidence-based case for investment in automated containment or improved segmentation, the consequence-centric counter to budgeting by fear.

3. **Benchmark Performance:** compare readiness for disruption against industry peers or against the organization's own history, which requires consistent self-assessment and surfaces real progress over time.

4. **Inform Strategic Decision-Making:** give leadership a clear, data-driven picture of the organization's exposure and its capacity to manage disruption, enabling better strategic choices than the threat-of-the-week could ever support.

Moving beyond compliance means measuring what truly matters: the ability of critical operations to survive and adapt. That demands a quantifiable understanding of resilience, so that the continuity of essential service is grounded in evidence rather than in the absence of a headline. But metrics, like controls, do not deploy themselves into a sector at scale. Whether a consequence-centric posture becomes the norm or remains the practice of a conscientious few is, in the end, a question of governance, and it is to the governance failures, and their remedies, that the chapter turns last.

The prevailing governance model resembles a patchwork quilt rather than a coherent tapestry. Energy, water, transportation, healthcare, and manufacturing each operate under disparate regulatory regimes, overseen by a constellation of agencies with their own priorities, methods, and enforcement. Some sector-specific tailoring is inevitable and even beneficial. But the absence of overarching strategic coherence produces gaps, redundancies, and inefficiencies, and it makes a unified national approach to operational continuity exceedingly hard to achieve. Information sharing is hampered by interagency friction; good practice fails to propagate across sectors; and the very definition of adequate security varies so widely that the result is a lowest-common-denominator effect in which proactive defense is stifled by the sheer complexity of navigating multiple, often conflicting, requirements. The deeper point for this chapter is that fragmented, compliance-driven governance entrenches exactly the wrong paradigm: it rewards the checking of boxes over the demonstrable capacity to withstand and recover.

The U.S. governance framework for OT cybersecurity is, in a word, fragmented. Reliant on a reactive, incident-driven posture, it has not kept pace with the conditions it is meant to address. The decentralization is the primary culprit. Responsibility is spread across numerous federal entities, including CISA, the Department of Energy (DOE), the Environmental Protection Agency (EPA), and the Department of Transportation (DOT), each with its own mandate and reporting requirements, and this diffusion of authority obstructs any coherent national strategy. Information sharing, the cornerstone of a stronger ecosystem, is choked by silos and competing priorities. Best practice develops slowly and unevenly as sectors adopt divergent readings of the same guidance, again producing the lowest-common-denominator effect in which genuine innovation is buried under the complexity of conflicting demands. The absence of a single authoritative voice leaves asset owners struggling to comply with a bewildering array of directives.

Compounding the structural problem is a pervasive culture of reactive policymaking, and here the connection to this book's thesis is exact. For decades, cybersecurity policy has been spurred by disruptive events. A major incident provokes urgent calls for new regulation, and the cycle of perpetual catchup resumes. Policy crafted under the pressure of the latest scare, without thorough analysis, broad consultation, or a real grasp of long-term OT implications, is policy built by the fear machine's tempo rather than by evidence. The reactive posture means security is bolted on after the fact rather than designed in, and the fear of the next attack consistently overshadows the strategic work of engineering intrinsic continuity, yielding short-term fixes that never touch the systemic weaknesses Chapter 8 laid bare.

The distinct nature of OT compounds the difficulty further. Unlike IT, which prioritizes the confidentiality of data, OT systems govern physical processes and prioritize safety, reliability, and continuous operation, with security historically a secondary consideration. Many legacy OT systems, designed for long lifespans in an era before pervasive cyber threats, simply lack modern security features, and their operational constraints, uninterrupted performance, proprietary protocols, the genuine hazards of patching a running process, make them ill-suited to off-the-shelf IT security practice. (The full structural account of why IT-derived solutions misfire in OT is given in Chapter 8; the point here is its governance consequence.) Governance must acknowledge and actively accommodate these differences rather than forcing OT into an IT-centric mold.

The disconnect between policy and reality is stark. Requirements written with IT in mind can mandate practices that are technically infeasible or economically prohibitive in OT. Frequent patching of a mission-critical control system can itself introduce instability and risk failure, directly contradicting the OT priority of availability, the very tension this chapter's treatment of patch management took pains to navigate. Stringent IT-style access policy can disrupt the seamless operation of an industrial process. And the absence of clear guidance on securing the legacy systems that remain prevalent across critical sectors leaves asset owners stranded between operational stability and compliance. The result is a culture of compliance theater, organizations meeting the letter of the law while achieving little real capacity to withstand disruption, which is precisely the failure mode a consequence-centric, metrics-based posture is designed to expose. The stakes are not abstract: critical infrastructure faces genuine risk, and a governance regime that produces paperwork instead of survivability leaves systemic weak points in place for any disruption, of any origin, to exploit.

Governance reform is therefore a national imperative, and it must move from a fragmented, fear-driven, IT-centric model toward one that is unified, proactive, OT-specific, and organized around demonstrable continuity rather than compliance. Several pillars define the shift.

Consolidate and Clarify Federal Authority

Complete centralization is infeasible, but federal agencies must coordinate far better than they do. Designating a primary agency, or standing up a robust, empowered interagency task force specifically for OT cybersecurity, could streamline effort, improve the flow of information, and produce a more cohesive national strategy. Such an entity would set overarching policy, coordinate incident response, facilitate the sharing of threat and failure intelligence alike, and oversee sector-specific guidance, leveraging existing agency strengths while establishing clear mandates and accountability. Clarity of authority is itself a continuity measure, because confusion about who is in charge is its own source of delay when a disruption strikes.

Shift from Reactive Policy to Proactive Resilience

Governance must actively incentivize and support a forward-looking, risk-based posture, which means moving past compliance checklists toward demonstrable capacity to withstand, adapt to, and recover from disruption. This is where the metrics developed earlier in this chapter become policy instruments rather than internal dashboards: the development of measures that accurately reflect OT security posture and operational continuity is the precondition for rewarding survivability instead of paperwork. Policy should encourage continuous risk assessment, scenario planning across the full spectrum of causes, and investment in security technologies genuinely suited to OT, rather than simply mandating a floor of minimum controls. Federal support for research and development into OT-specific solutions matters too, fostering innovation in secure industrial communication protocols, anomaly detection for industrial processes, and resilient control-system architectures, the engineered robustness on which everything in this chapter depends.

Adapt Governance to OT-Specific Realities

Governance frameworks must be designed explicitly around the operational realities of industrial control systems: the primacy of availability and safety, the challenges of legacy equipment, and the specific failure modes OT environments actually face. Rather than retrofitting IT security solutions, policy should promote OT-native approaches, robust network segmentation, defense-in-depth adapted for industrial settings, and secure remote access that prioritizes operational integrity. Close collaboration with OT vendors and asset owners is essential to ensure the resulting policy is practical, implementable, and genuinely effective rather than another layer of theater.

Address the Critical Workforce Gap

The shortage of specialized OT cybersecurity skills cripples collective defense, and it is, in the framing of this chapter, a continuity risk in its own right: a sector cannot withstand and recover from disruption it lacks the people to manage. Reform must mandate substantial investment in education, training, and workforce development, federal funding for university programs, apprenticeships, industry-led training, and public awareness campaigns to draw talent into an essential field. Retaining skilled personnel within critical-infrastructure sectors is equally vital, and clear career pathways and professional development are the means of building and sustaining the expertise.

Foster Public-Private Collaboration and Information Sharing

Because the vast majority of critical infrastructure is privately owned, effective collaboration between government and industry is essential. It requires building trust and creating secure mechanisms for sharing both threat intelligence and the hard-won lessons of failure, the very exchange a fear-driven, secrecy-prone culture suppresses. Existing public-private partnerships and sector-specific ISACs must be strengthened, and government guidance must be developed in close consultation with industry. Incentivizing voluntary information sharing through clear legal protections and demonstrated benefit is the way to dismantle the hesitations that the threat-centric, target-on-my-back mindset has bred.

Reform Procurement for Long-Term Resilience

Procurement reform is essential to sustainable security. Government and industry must shift from acquiring static point solutions toward investing in adaptable, long-lived capacity for continuity. That means mandating interoperability to avoid the proprietary silos that create blind spots and hinder coordinated response. It means addressing the entire system lifecycle, requiring vendors to publish transparent end-of-life (EOL) and end-of-support (EOS) policies and to commit to long-term security patching, since a system that outlives its support is a fragility waiting to be realized. And it means making procurement cycles more agile, so that modern technology can be acquired before it is obsolete rather than long after.

An Action Plan for OT Security Reform

To translate these principles into action, the following roadmap outlines specific initiatives across three timelines. Each is framed not as a defense against a named adversary but as a step toward a national infrastructure that can withstand and recover from disruption whatever its cause.

Within 90 Days: Foundational Actions

Initiative	Owner(s)	Key Performance Indicators (KPIs)	Action Type
Clarify Federal OT Security Roles	White House (NSC), OMB	National Security Executive Memorandum (NSM) published clarifying agency roles (CISA, DOE, EPA, DoD, etc.) and designating a lead federal entity for OT security coordination.	Executive
Launch Public-Private OT Supply Chain Task Force	CISA, NIST, Dept. of Commerce	Task force chartered with public and private sector members. Initial report on high-risk components delivered to CISA Director.	Executive
Issue OT Security Guidance for SLTTs	CISA, FEMA	Joint guidance published for State, Local, Tribal, and Territorial governments on accessing federal resources and grants for OT security.	Executive

On the ground, these initiatives provide immediate clarity. Asset owners will have a single, clear point of contact for incident reporting, procurement decisions will be informed by a federally recognized list of high-risk components, and local governments will have direct guidance for accessing funds to purchase new monitoring sensors or conduct security audits.

Within 1 Year: Building Capacity

Initiative	Owner(s)	Key Performance Indicators (KPIs)	Action Type
Mandate SBOM for Federal OT Procurement	OMB, CISA	Final rule issued requiring a Software Bill of Materials (SBOM) for all new OT systems acquired by federal agencies.	Executive

Initiative	Owner(s)	Key Performance Indicators (KPIs)	Action Type
Establish OT Resilience Grant Program	CISA, FEMA	Grant program established and first application window opened for SLTT and critical infrastructure operators. $50M in grants awarded.	Legislative
Develop Standardized OT Resilience Metrics	NIST, CISA	NIST Interagency Report (NISTIR) published defining core OT resilience metrics (e.g., MTTD, MTTR, RTO) for voluntary adoption and use in grant programs.	Executive
Pilot USCYBERCOM Managed Defense for a CI Sector	DoD (USCYBERCOM), CISA, Sector Regulator (e.g., DOE)	Pilot program established for one subsector (e.g., bulk electric system). Rules of engagement defined. Report on initial findings delivered to Congress.	Executive

On the ground, these actions build tangible capacity. Federal asset owners gain authority to reject systems with insecure software, critical infrastructure operators can access new grants to upgrade legacy systems or install new sensors, and voluntary resilience metrics create standardized reporting thresholds for audits, the first concrete instance of governance rewarding measured survivability rather than box-checking.

This voluntary, opt-in pilot would be offered to a critical infrastructure subsector based on national importance and willingness to collaborate. 'Managed defense' includes network monitoring, threat hunting, and incident response support from USCYBERCOM, operating under civilian oversight from CISA and the relevant regulator. Clear rules of engagement would protect privacy and prevent operational disruption. The goal is to test a defense model against nation-state threats, not to militarize domestic infrastructure.

Within 3-5 Years: Strategic Realignment

Initiative	Owner(s)	Key Performance Indicators (KPIs)	Action Type
Pass National OT Vendor Security & Liability Law	Congress	Legislation passed establishing secure-by-design principles, vulnerability disclosure requirements, and a safe harbor liability framework for OT vendors.	Legislative

Initiative	Owner(s)	Key Performance Indicators (KPIs)	Action Type
Fund National OT Workforce Development Centers	Dept. of Education, NSF, Dept. of Labor	At least five National Centers of Excellence for OT Cybersecurity established via federal grants to university consortia. 5,000 professionals certified.	Legislative
Modernize All Sector-Specific Regulations	Sector Risk Management Agencies (SRMAs)	All major SRMA regulations updated to incorporate resilience-based standards from the NISTIR and require regular, independent OT security audits.	Executive

On the ground, this strategic realignment fundamentally changes the ecosystem. New laws will shift the security burden to vendors, national centers will produce a workforce capable of performing specialized audits, and modernized regulations will replace outdated checklists with mandatory, resilience-based security assessments, the consequence-centric paradigm written, at last, into the rules themselves.

The concept of long-term resilience also demands a shift in how cybersecurity solutions are evaluated during procurement. Many OT systems have lifecycles measured in decades, not years, and a solution bought today must be capable of evolving with the conditions of tomorrow and of being supported across its entire operational lifespan, because a system that outlives its vendor's commitment becomes, by definition, a fragility no one is maintaining. Vendors should therefore be evaluated not only on a product's immediate capabilities but on their commitment to ongoing support, regular updates, and proactive vulnerability management. Procurement policy should weigh the availability of long-term maintenance agreements, the vendor's track record for timely patch releases, and their commitment to secure software development lifecycle (SDLC) practices. A solution that is state-of-the-art today but lacks a clear roadmap for future development and support is a significant long-term risk to continuity, however impressive its current feature list.

The procurement process itself must become more agile. Traditional, lengthy procurement cycles can deliver solutions already outdated by the time they are deployed. Agencies need the flexibility to fold emerging threats and technologies into their procurement strategy, which may mean adopting more agile methodologies such as iterative development and phased acquisition, allowing continuous evaluation and adjustment of requirements and solutions. Government can also help by establishing and promoting standardized security frameworks and architectures for OT, which then guide procurement toward consistency and interoperability across sectors. A government-developed "Secure OT Architecture Blueprint," for example, could give agencies and operators a standardized template against which to evaluate and procure solutions aligned with sound practice for segmentation, access control, and monitoring, embedding the foundational controls of this chapter into the act of purchase.

Closing the workforce gap begins with strengthening educational pathways. Universities and vocational schools need support to develop specialized curricula that bridge IT security and OT principles, covering industrial protocols, ICS architecture, and embedded systems, with industry partnerships providing internships and keeping the programs relevant.

Beyond academia, the existing workforce must be upskilled. Engineers, technicians, and plant managers with deep domain expertise need targeted training to identify and mitigate cyber risk, training that converts operational experts into invaluable security assets, the very people best positioned to recognize a system beginning to degrade and to act before it fails.

Fostering a culture of security awareness across every level of an organization is paramount, and it is the cultural argument of this chapter extended to the national scale. Security cannot be the sole province of a dedicated IT or security team; it must be ingrained in the operational ethos of everyone who touches an OT system, through continuous reinforcement of good practice, regular communication about emerging risk, and clear mechanisms for reporting suspicious or anomalous activity. Leadership must champion this culture, demonstrating a visible commitment and allocating the resources for training and awareness. When security is a shared responsibility, the collective vigilance of the workforce becomes a powerful deterrent, and, just as important, an early warning system against failures of every kind.

Investing in workforce development is, finally, an investment in resilience. The cost of a successful disruption of critical infrastructure, in downtime, in fines, in physical consequence, is orders of magnitude greater than the cost of training, and equipping people with the right skills both reduces the likelihood of incidents and shrinks their impact when they come. The recruitment of talent into the OT sector is itself a challenge, since the higher salaries and more established career paths of IT cybersecurity tend to pull talent away. The answer is to make plain the unique and consequential nature of the work: protecting power grids, water treatment plants, and transportation networks offers a sense of purpose, a direct contribution to the public good, that can be a powerful draw for skilled professionals, reinforced by clear career progression, competitive compensation, and opportunities for advanced specialization.

The development of simulation and training platforms purpose-built for OT is a critical area for investment as well, and it is the natural culmination of everything this chapter has argued. Such platforms replicate the complexity of industrial control systems without the risk of experimenting on live ones, letting personnel practice incident response, test defensive strategies, and build hands-on experience in a safe, controlled environment, rehearsing not only the cyberattack but the equipment failure, the environmental shock, and the cascading fault. They are invaluable for developing the muscle memory and decision-making that high-pressure incidents demand, the human capacity to withstand and recover that no technology can supply on its own. Government funding and public-private partnership can accelerate the development and adoption of these technologies, and in doing so accelerate the shift this chapter has argued for throughout: away from the fear machine's endless, brittle, ruinously expensive pursuit of every imaginable adversary, and toward the durable, evidence-based, and ultimately more secure work of building systems, and people, that bend without breaking and recover before the harm is done.

CHAPTER 14

CONCLUSION: SECURING AMERICA'S FUTURE IN A COMPLEX WORLD

The preceding chapters have meticulously dismantled the prevailing "Cybersecurity Fear Machine" that dominates the discourse surrounding Operational Technology (OT) security in the United States. This machine of sensationalism, vested interests, and policy inertia has consistently amplified the perceived threat to our nation's critical infrastructure eclipsing verifiable evidence. We have navigated the landscape of OT vulnerabilities, opaque threat assessments, and the forces that perpetuate a crisis atmosphere. This chapter does not re-argue that case from the beginning. It consolidates what the evidence has shown, states the argument once in its strongest form, and then turns fully toward the path forward.

Stated plainly and stated once: the narrative of impending OT catastrophe has been significantly overblown, and the machine that produces that narrative is neither accidental nor benign. Cyber threats to critical infrastructure are real. They are not, on the available evidence, the imminent, civilization-ending peril that the dominant discourse describes. Between that real-but-bounded risk and the manufactured perception of perpetual doom sits an entire apparatus of incentives, and this book has been an effort to name its parts. The commercial sector profits when danger is assumed to be escalating and existential. Fragmented governance, lacking a unified risk-based strategy, defaults to reaction, and reaction is fertile soil for fear. The long lifecycles and genuine complexity of OT systems are presented not as engineering conditions to be managed but as proof of inherent fragility. And the historical record, examined without flinching, does not support the story of routine, nation-altering collapse. That is the whole of the thesis. The remainder of this chapter assumes it rather than relitigates it.

It is worth being precise about what this argument is and is not, because the temptation to overstate runs in both directions. The claim is not that OT cannot be attacked, nor that the consequences of a successful attack are trivial, nor that vigilance is wasted. The claim is narrower and more demanding: that the industry systematically overstates the evidence, that rare events are routinely mislabeled as typical ones, and that this inflation is itself a cost borne by the very infrastructure it claims to protect. A book whose central complaint is the overstatement of evidence cannot end by overstating the opposite case. So the conclusion holds the line exactly where the data holds it: the threat is real but inflated; catastrophic physical-consequence attacks are rare, not impossible; and the rational response is to redirect resources from fear-driven theater toward evidence-based resilience.

The Machine's Two Engines, Named Once

If the body of this book has a single contribution beyond cataloguing incidents, it is in identifying what now powers the machine, because the power source has changed and most of the discourse has not caught up to that change. More than a decade ago, in 2011, Brito and Watkins warned of a "cyber-industrial complex" — a critique aimed squarely at government, at the way appropriations and contractor relationships and threat briefings could feed one another and inflate spending. That warning was correct for its moment, and it remains a useful lens. But it is no longer the whole picture, and treating it as the whole picture lets the most important development of the last decade pass unexamined.[1] The original contribution of this book is to update that 2011 framing into its 2020s form, which is no longer principally about public budgets at all. The threat narrative has been financialized. It is now owned, packaged, and traded as private capital. Private equity and venture capital have discovered that fear, properly maintained, is not merely a sales technique but an appreciating asset class. The story of escalating, existential OT danger underwrites valuations; the valuations create constituencies with a direct financial interest in keeping the story alive; and the cycle compounds. The scale of that appreciation is not abstract. A single security firm's valuation rose from roughly $1.1 billion to an agreed sale at $7.75 billion over the span of five years

— built on the same fear thesis, with the underlying record of demonstrated physical harm essentially unchanged across that period.[2] That is the machine's first and most powerful engine, and Chapter 4 develops it in full. Here it need only be named: where Brito and Watkins saw a complex of government spending, we now confront a complex of private capital, and the narrative itself has become the asset on the balance sheet.

The distinction is not academic, and it is worth dwelling on for a moment because it changes what reform must target. A government spending complex, for all its inertia, is at least answerable in principle to the ordinary instruments of public accountability — appropriations hearings, inspectors general, the eventual electoral consequence of waste. The threat narrative could in theory be deflated by a sufficiently determined legislature or a sufficiently skeptical oversight committee, because the money flowed through channels the public owned. Private capital answers to none of that. Its discipline is the return, and the return on a financialized fear thesis improves precisely as the fear is sustained and broadened. When a narrative becomes an appreciating asset, the people who hold it are not merely indifferent to deflation; they are structurally opposed to it, because deflation is a write-down. The incentives that once pushed in the direction of inflated public budgets now push in the direction of an inflated story carried on private balance sheets, defended by parties whose fiduciary obligation is to protect the valuation — which is to say, to protect the fear. This is why the older critique, accurate as it was, is insufficient on its own: it diagnosed a complex that could be reached by public levers, and the complex has since moved somewhere those levers do not reach. Recognizing that migration is the precondition for any honest reform, because a remedy aimed at the 2011 target will miss the 2020s one entirely.

The second engine is governance, and it has been mischaracterized for so long that the mischaracterization deserves a final correction. Responsibility for securing critical infrastructure is spread across numerous federal agencies, state governments, and regulatory bodies, each with its own mandates, priorities, and reporting mechanisms. This diffusion of authority produces a reactive rather than proactive posture: when incidents occur, the response is new directives, advisories, or regulations, often without a comprehensive, cross-sectoral strategy. The conventional reading treats this as a resourcing problem — not enough money, not enough tools, not enough attention. The reading this book has argued for is different and more uncomfortable. Fragmentation does not merely leave gaps; it creates a vacuum that fear is well suited to fill, because in the absence of a grounded, risk-based strategy that ranks vulnerabilities by objective data, the discourse defaults to generalized anxiety about unknown, potentially devastating, future attacks. Fragmentation also creates a market. It opens space for actors to lobby for mandates that favor particular technologies or services rather than performance-based outcomes, embedding the fear narrative directly into policy. The two engines, in other words, are coupled: private capital supplies the narrative, and fragmented governance supplies the conditions under which the narrative cannot be effectively challenged or measured against results.

What the Record Actually Shows

Against the volume and confidence of the alarmist account, the empirical record is strikingly thin, and the thinness is the single most important fact in this book. It is worth stating in the plainest possible terms. Roughly fifteen years have passed since Stuxnet — the event that inaugurated the modern era of OT cyber-fear and remains its founding reference. In that decade and a half, the independently verified list of cyberattacks that produced confirmed cyber-physical effects numbers about four.[3] Not four per year. Four, total, across fifteen years and across the entire global landscape of industrial control systems. The sources for this chapter, gathered below, trace the canonical members of that short list, and a reader can examine each on its own terms.

Set that number beside the other side of the ledger. Information technology breaches — the ordinary, nonphysical compromises of data and networks — run to tens of thousands of incidents every year.[4] The asymmetry is not a rounding error or a matter of definitional taste. It is four orders of magnitude, and it is the gap the fear machine works hardest to obscure, because the machine's basic rhetorical move is to borrow the frequency of IT incidents and lend it to the consequences of OT catastrophe — to take the tens of thousands of breaches that happen and imply that any one of them could be the centrifuge, the blast furnace, the darkened grid. The historical data, examined objectively, does not sustain that implication.

The discipline that produces the short list is itself part of the argument. The figure of "about four" is not offered in the same currency as the industry's projections. Counting an event requires that the incident be independently verified, that the effect be cyber-physical rather than merely informational, and that the cause be traceable to cyber action rather than coincidence, weather, wear, or human error misfiled after the fact as an attack. Apply that test honestly and candidates thin out quickly: Stuxnet physically degraded centrifuges; the two Ukraine grid attacks caused bounded service disruptions; TRITON targeted a safety system and was discovered after a fail-safe trip;

FrostyGoop showed a bounded ICS-driven heating outage; and cases such as the German steel-mill report or PIPEDREAM/Incontroller belong in separate disputed or capability categories. The smallness of the set is not a failure of diligence; it is the product of diligence — it is what remains once the inflation is stripped away. The contrast with the IT ledger is instructive for the same reason: the identical demanding test, applied to ordinary network breaches, returns tens of thousands of confirmed events every year. The method is not rigged to produce a small number; it produces a small number for OT cyber-physical effects and an enormous one for IT compromise, which is precisely what an honest accounting of two genuinely different phenomena ought to do.

This is also where rigor cuts against the author as much as against the industry. The thinness of the confirmed cyber-physical record does not prove that OT systems are safe, and it must not be read that way. Some of the four incidents were genuinely serious; the conditions that produced them were specific and rare, not impossible; and the absence of a fifth or a fiftieth is not a guarantee about the future. What the record supports is a claim about proportion, not a claim about immunity. Many reported "OT incidents" turn out, on inspection, to have been contained, localized, or rooted in human error, physical security lapses, or routine operational failure rather than sophisticated nation-state cyber-warfare. The leap from a detected intrusion to a nationwide shutdown of essential services is enormous and requires a rare confluence of conditions. The machine erases that distance, equating attempted intrusions and minor disruptions with imminent collapse, and in doing so it does real harm: it channels investment toward hypothetical worst cases and away from the unglamorous foundations — asset inventory, patching disciplined to operational constraints, rehearsed incident response — that actually determine whether a plant stays running.

The Real Vulnerability Is Not in the Code

If the threat has been inflated, it does not follow that there is nothing to fear. It follows that we have been afraid of the wrong thing. The nation's true vulnerabilities, as this book has endeavored to demonstrate, do not reside primarily in the sophistication of adversarial cyber capability, real and persistent though that is. They reside in the systemic weaknesses of the American governance and institutional frameworks charged with protecting critical infrastructure. The pervasive narrative of imminent catastrophe has functioned as a powerful distraction, drawing attention and resources away from the patient work of repairing those weaknesses.

The architecture of our national security apparatus was designed for a different era and strains against the dynamic character of modern threats to operational technology. Fragmented responsibility breeds inertia; coordinated action turns sluggish; agencies operating with distinct priorities and limited information sharing produce a patchwork of policy that fails to build systemic resilience. The exposure that results is not, in the main, exposure to the most advanced adversary. It is exposure to persistent, ordinary failure — the failure that comes from a lack of integrated planning, interoperable systems, and a well-trained workforce.

This institutional condition is itself the critical vulnerability. In OT environments, where systems are built for long lifecycles and uptime is paramount, the pace of technological change collides with security imperatives, and the well-founded reluctance to disturb a running control system hardens into resistance that mandates struggle to overcome. That reluctance is frequently a rational response to physical risk; coupled with fragmented oversight, however, it becomes a dangerous stasis. Absent clear, consistent, actionable guidance that accounts for OT's operational constraints — and absent real incentives to modernize — the security posture of critical sectors stays anemic. The fear machine offers a tempting and superficial remedy: buy more tools. That remedy never reaches the underlying problem, which is the absence of a strategy for resilience encompassing modernization, workforce development, and adaptive governance. The true deficit is an inability to marshal resources, align stakeholders, and sustain long-term programs — not a shortage of security appliances to install in response to manufactured crises. And each manufactured crisis that fails to arrive spends down the credibility the next, possibly genuine, warning will require — the machine's final irony, that in selling vigilance it quietly corrodes the public's capacity for it.

Resilience, properly understood, is the capacity to absorb a disruption, adapt to it, and recover quickly, and it cannot be bought in a box. It requires a holistic posture spanning cyber defense, physical security, supply-chain robustness, personnel preparedness, and operational continuity. The fragmented model actively obstructs that posture, leaving each sector and often each organization to fend for itself with a contradictory stack of advisories, so that capability is distributed unevenly — a few well-defended operators alongside many that remain acutely exposed. By fixating on the specter of the sophisticated attack, the machine diverts attention from the prosaic, equally dangerous failures that follow from incoherent planning: the absence of clear incident-response authority across agencies that turns a minor event into a confused and delayed one; the lack of unified standards for intelligence sharing that hobbles both preparation and response.

These are the cracks in the foundation, and no quantity of fear-driven procurement fills them.

What Evidence-Based OT Policy Looks Like

The argument of this book has been corrective and largely diagnostic. It would be incomplete, and a little cowardly, to stop at diagnosis. So it is worth saying concretely what the alternative is — what policy looks like when fear is removed as its organizing principle and evidence is installed in its place.

The governing shift is from a threat-centric to a consequence-centric stance. Threat-centric policy begins with the adversary — its presumed capabilities, its imagined intentions, the worst it might conceivably do — and works backward, which is precisely the orientation that lets a four-incident record underwrite a tens-of-thousands-incident climate of dread, because the adversary's hypothetical ceiling is unbounded and therefore inexhaustibly marketable. Consequence-centric policy begins instead with the system and asks a disciplined, answerable question: which functions, if interrupted, actually matter, how badly, for how long, and to whom? It ranks critical processes by the real-world harm of their failure, maps the dependencies among them, and concentrates resources where interruption would be genuinely intolerable. It treats the assurance of core functions, not the prevention of every intrusion, as the objective — because preventing every intrusion is an unwinnable, and therefore endlessly fundable, war, whereas assuring continuity is a finite and measurable goal. This is the resilience paradigm developed in Chapter 13, carried here to its policy conclusion: stop measuring security by the height of the wall and start measuring it by how quickly the essential keeps running after the wall is breached.[5]

The difference is easiest to see in the concrete. Consider how the two approaches handle the same plant. A threat-centric review opens with the adversary catalogue: which nation-states and criminal groups are said to target this sector, what tooling they are reported to possess, what the most advanced of them might in principle accomplish. It then shops for products that promise to counter each item on that catalogue, and because the catalogue is open-ended — there is always a more capable hypothetical adversary, always a newer reported technique — the shopping never ends and the budget tracks the catalogue rather than the plant. A consequence-centric review of the very same facility begins at the other end entirely. It asks the engineers, not the threat-intelligence vendors, a sequence of plain questions: which processes here, if they stopped or ran wrong, would actually hurt someone, foul the environment, or halt production, and which would merely be inconvenient? Of the processes that matter, which depend on which others, so that a failure in one cascades into the rest? Where are the points at which a loss of control would be hardest to detect or slowest to reverse? Having answered those questions, the review concentrates effort on the handful of functions whose interruption is genuinely intolerable and accepts a measured, documented level of risk on the rest — not because the rest cannot be attacked, but because finite resources spent everywhere protect nothing in particular. The threat-centric plant ends up with an expensive, sprawling defensive perimeter and no clear answer to the only question that matters when something goes wrong: will the essential process keep running, and if not, how fast can we bring it back? The consequence-centric plant is organized around that question from the first day. One posture is built to survive the audit and the sales call; the other is built to survive the incident. The asymmetry the empirical record exposes — roughly four category-dependent cyber-physical cases against tens of thousands of ordinary breaches — is itself an argument for the consequence-centric orientation, because it tells us with unusual clarity which consequences are rare and grave and which are common and survivable, and a rational allocation of a fixed budget follows directly from that ranking.

Such a posture changes what gets measured, and what gets measured changes what gets funded. Compliance checklists give way to performance-based outcomes expressed in terms an operator can verify and an overseer can audit: mean time to recover for a critical process; the completeness of asset inventory; the demonstrated, exercised ability to maintain or restore essential service under degraded conditions. These metrics have the singular virtue of being indifferent to the press release. A vendor's threat report, however breathless, cannot move them; only real improvement in real resilience can. That indifference is exactly the property a fear-saturated field most needs, because it severs the link between the volume of the alarm and the direction of the spending.

It is worth being concrete about what such measurement does to a security program in practice, because the change is not cosmetic. Under a threat-centric, compliance-driven regime, success is a binder: the controls were documented, the boxes were checked, the audit was passed, and none of it tells anyone whether the plant would actually keep running. Under a consequence-centric, performance-driven regime, success is a demonstration: the team ran the drill, the critical process was restored within the target window, and the gap between target and result is a number that can be tracked quarter over quarter. Mean time to recover, in particular, has the useful property of being hard to fake — either the process came back in the time claimed or it did not — and of pointing investment toward the things that genuinely shorten it, which are almost never the most expensive items on a vendor's price sheet. A complete and current asset inventory shortens it, because one cannot recover what one cannot find. A tested backup shortens it; an untested one merely promises to. A rehearsed response plan shortens it; a shelved one does not. The metric, in other words, quietly redirects attention to exactly the unglamorous foundations the fear machine teaches operators to neglect, and it does so without anyone having to win an argument about how frightening the latest threat report is. The number simply does not care, and that indifference is the whole point.

Evidence-based policy also requires oversight built to match. The oversight model set out in Chapter 12 is the institutional complement to the resilience paradigm: independent assessment that can distinguish a demonstrated risk from a marketed one, that can hold both vendors and agencies to outcomes rather than activity, and that can break the feedback loop in which the parties profiting from a threat are also among its loudest narrators.[6] Without that independent vantage, performance metrics are eventually captured and redefined by the same incentives that inflated the threat in the first place; with it, the financialized narrative meets a counterweight whose interest is accuracy rather than appreciation. Policy, metrics, and oversight thus form a single structure. Consequence-centric policy decides what to protect and why; performance-based metrics determine whether the protection works; independent oversight keeps both honest. Remove any one and the other two drift back toward theater.

None of this is a retreat from security. It is a more demanding version of it. Fear is easy; it requires only a budget and a worst case. Evidence is hard; it requires inventories that are tedious to compile, drills that are inconvenient to run, metrics that are unflattering to report, and the discipline to spend against documented risk rather than against the morning's headline. The reward for that difficulty is an infrastructure whose strength is real and demonstrable rather than asserted — built on the bedrock of evidence rather than the shifting sands of manufactured anxiety.

The Path Forward: Federated Governance and Operational Resilience

Diagnosis and principle still leave open the question of structure: through what institutional form is consequence-centric, evidence-based security actually delivered across a landscape as varied as American critical infrastructure? No single government or private entity holds the authority or the understanding to manage the whole diverse terrain of OT security, and any model that pretends otherwise simply recreates the fragmentation under a new name. The answer this book proposes is a federated governance model wedded to an unwavering commitment to operational resilience — distributed responsibility under genuine coordination, paired with a relentless focus on the continuity of essential service. It is a practical necessity, not a theoretical ornament, and it is the organizational expression of everything argued above.

A federated model empowers sector-specific coordinating bodies that bring industry leaders, government regulators, and technical experts to the same table for energy, water, transportation, manufacturing, and the rest. These bodies become the working nodes for threat-intelligence exchange and for best practices tailored to each sector's actual operating environment, replacing generalized pronouncements of danger with specific, context-aware risk assessment — the same move from abstraction to particulars that consequence-centric policy demands. They must be underpinned by robust systems for collecting, analyzing, and disseminating intelligence, the object of which is not the trading of raw data but the cultivation of a shared, pattern-level understanding of the threat landscape, achieved through trusted information-sharing communities where proprietary concern is balanced against national interest under standardized protocols for classification, anonymization, and secure communication. Built into the structure must be mechanisms for continuous feedback and improvement, so that processes are refined against the lessons of real incidents and exercises, and so that the connection between cyber and physical security is bridged rather than left to fall between disciplines. The point of the architecture is not coordination for its own sake; it is the situational awareness that makes a swift, unconfused, well-rehearsed response possible when something actually happens.

Resilience is the other half of the structure, and it must be read broadly rather than narrowly. It is far more than the technical capacity to repel intrusions. It encompasses the entire ecosystem — physical security, supply-chain integrity, skilled personnel, rehearsed continuity plans — and it lives in technology, in process, and in people at once. Technologically it means redundant systems, designs that degrade gracefully rather than failing catastrophically, and backup and recovery that have been tested under realistic conditions. In process it means business-continuity planning, regular drills, and a standing culture of preparedness. In people it means investment in a skilled OT cybersecurity workforce fluent in both industrial control systems and the evolving threat landscape, and a security-aware culture reaching from the plant floor to the executive suite. This is where the evidence-based approach earns its keep in dollars: rather than pouring capital into speculative, high-cost solutions sold on fear, resources flow to practical, proven measures that improve the ability to withstand and recover — rigorous asset management so that every component of the OT environment is known, patching and vulnerability programs that respect operational constraints, incident-response protocols tested often enough to be trusted. The federated model and operational resilience reinforce each other directly. A federation that brings OEMs, asset owners, and regulators into the same room can build shared roadmaps for modernizing legacy systems and for designing future deployments to be resilient from the start, overcoming the inertia that no single actor can overcome alone — and it can do so because all parties come to understand that collective security and collective resilience are one and the same. A disruption in the energy sector can cascade into water treatment, transportation, and communications; only a coordinated, cross-sectoral risk assessment can see those dependencies before they fail, and only a federated structure can convene one. This embrace of a federated structure and a resilience-first posture is not complacency and not an underestimation of the threat. It is the assertion that the prevailing approach is suboptimal — inefficient, misallocated, and finally unequal to the enduring resilience the nation's vital systems require — and that a better one is within reach for those willing to shed an outdated paradigm in favor of evidence.

A fair objection arises here, and it should be met rather than sidestepped: federation of this kind cannot simply be decreed. A statute can announce a coordinating body, but it cannot legislate the trust on which information sharing actually depends, and without that trust the body becomes one more layer of the fragmentation it was meant to cure. The transition therefore has to be sequenced, and the sequence runs from the bottom upward rather than the top down. It begins, as the action steps that close this chapter suggest, with small cells of operators who already have reason to trust one another — peers in the same sector, often the same region, who face the same equipment, the same suppliers, and the same operational constraints. Such a cell can start exchanging practical observations long before any formal charter exists: a strange firmware update from a shared vendor, an anomaly seen on a common control platform, a defensive tactic that worked. As these cells demonstrate their worth, they accrete. Trust earned at the working level becomes the substrate on which sector-wide coordination can later be built, because by then the participants hold a record of reciprocity rather than an order to comply. Government's most useful role in this sequence is not to command the structure into existence but to lower the barriers to its formation — supplying the standardized protocols for classification and anonymization that let competitors share safely, offering the legal protections that keep a candid disclosure from becoming a liability, and resisting the temptation to convert every voluntary exchange into a reporting obligation that would extinguish the candor in the first place. Built this way, the federation is durable because its participants own it; built the other way, by fiat, it inherits the very inertia and mistrust that crippled the fragmented model. The patience this demands is itself a departure from the fear machine's habits, which always prefer the dramatic announcement to the slow accumulation of working relationships — which is precisely why the slow accumulation is the more credible path.

From Theory to Action

Moving from diagnosis to action requires concrete steps. For leaders in government and critical infrastructure, these principles translate into four priorities:

1. **Mandate an Evidence-Based Risk Review.** Reevaluate your top five OT cyber risks using the rubric from Chapter 2. Compare this assessment against current security spending and priorities. Disconnect budget from headlines and reconnect it to documented risk.

2. **Prioritize Resilience Over Compliance.** Shift the focus from box-checking compliance to demonstrable resilience. Launch a pilot project to measure your team's Mean Time to Recover (MTTR) for a critical OT process and make reducing that time your primary security goal for the next quarter.

3. **Establish a Sector-Specific Intelligence Cell.** Instead of passively consuming generic threat feeds, create a small, federated cell with trusted peers. The goal is to share practical observations on operational anomalies, supply chain issues, and defensive tactics, building the foundation for the federated model.

4. **Redirect One Major Security Purchase.** Veto one significant, fear-driven procurement of a "next-generation" tool. Reallocate those funds to foundational projects: completing an OT asset inventory, funding a patching program for a legacy system, or improving physical security.

Securing America's future is not about buying more technology to fight a hypothetical cyber war. It is about the disciplined, often unglamorous work of strengthening governance, building operational resilience, and grounding our security strategy in evidence rather than anxiety. Each of the four steps above is small on its own, and deliberately so. The machine was not built in a single decision and it will not be dismantled in one; it is sustained by a thousand individually reasonable choices to spend against the worst imaginable case, and it is undone by an equal number of choices to spend against the documented one instead. None of these four steps requires waiting for federal reorganization or for a new appropriation. Each can begin inside a single organization, this quarter, with resources already in hand — which is the point. The shift from fear to evidence is not, in the end, a posture a nation adopts by proclamation. It is the accumulated result of operators and officials choosing, case by documented case, to act on what the record shows.

The shift toward evidence-based policymaking is not a call for reduced security, but for a more intelligent and effective application of those efforts. It is about ensuring that our investments in OT cybersecurity are guided by facts, not fear. It is about building a national security strategy that is resilient, adaptable, and grounded in the realities of our interconnected industrial world. This commitment to data, transparency, and objective risk assessment is the bedrock upon which a truly secure future for America's critical infrastructure must be built. Without it, we risk expending vast resources on strategies that fail to address the most pressing threats, leaving our nation vulnerable. The imperative is clear: to secure America's future, we must first understand it, and understanding begins with evidence.

A realistic strategy demands an honest appraisal of the capabilities and limits of current cybersecurity practice within the particular context of OT. Unlike information-technology environments, which are generally designed with security and upgradeability in mind, OT systems are built for reliability, availability, and longevity, and downtime for patching or upgrade can carry immediate, severe consequences for physical processes — risks to public safety, environmental integrity, and economic output. That operational reality means security solutions must be tailored to OT's constraints: the "move fast and break things" reflex of the IT world is antithetical to OT's core principles, so defenses must be engineered for minimal disruption, rigorous testing, and deep integration with operational workflow. The goal, once more, is not absolute security — an unattainable ideal whose pursuit is precisely what the fear machine sells — but the effective management of risk and the assurance of continuity, which means understanding the cascading effects of failure and building the means to contain and recover from incidents quickly.

The machine also habitually overlooks the human factor, and a balanced strategy cannot. Technology supplies the tools, but people design, implement, operate, and maintain the systems, and so robust training, clear operational procedures, and a genuine security culture matter as much as any appliance. That includes educating personnel on phishing and social engineering and on the discipline of following protocol; it includes building an environment where employees can report suspicious activity without fear of reprisal; and it includes confronting the sector's persistent skills gap through education, apprenticeship, and knowledge-sharing, so that the specialized expertise OT security demands is actually available where it is needed. Security-awareness programs written for the real roles and responsibilities of OT personnel mitigate a category of risk that no technology addresses on its own.

The economics belong in the equation as well. Comprehensive security is expensive, especially for smaller operators and those on thin margins, and a fear-driven posture that reaches reflexively for the most expensive, most advanced solution produces a security gap in which only the largest and best-resourced organizations can afford adequate protection. A realistic strategy is tiered, prioritizing investment by risk assessment and by the actual impact of a compromise — leveraging open-source tooling where it fits, securing foundational practice first, and providing public support or incentives so that critical infrastructure entities can adopt the measures that matter. The principle of security by design must be woven through the OT lifecycle, pressing manufacturers to build security in from the outset rather than bolt it on as an afterthought.

The "Cybersecurity Fear Machine" served a purpose: in its early years it raised awareness of genuine vulnerabilities that had gone too long ignored. That contribution should be acknowledged honestly, because the argument of this book is not that the threat is fictional. The argument is that the machine has outlived its usefulness, that its continued operation now paralyzes progress and misdirects the very resources it claims to protect, and that recognizing its mechanics is the first step toward replacing it with something better. Securing America's future in a complex world will not be won as an arms race against an ever-evolving roster of adversaries. It will be won, if it is won, through the robust, resilient, and adaptive governance of the systems on which our society depends. The true vulnerabilities lie not in the code but in the interaction of policy, people, and process that underpins the nation's critical infrastructure. Overcoming the inertia of fragmented governance and dismantling the fear machine are the essential first steps toward a future in which our infrastructure is defined not by its advertised fragility but by its demonstrable strength. The path forward demands a turn away from the obsession with hypothetical doomsday and toward a pragmatic, persistent focus on the foundations of governance and resilience — so that America's future rests on the bedrock of evidence, and not on the shifting sands of manufactured fear.

Sources for the Conclusion

1. Nicolas Falliere, Liam O'Murchu, and Eric Chien, "W32.Stuxnet Dossier," version 1.4 (Symantec Security Response, February 2011), via the National Security Archive, `https://nsarchive2.gwu.edu/NSAEBB/NSAEBB424/docs/Cyber-044.pdf`; and Ralph Langner, "To Kill a Centrifuge: A Technical Analysis of What Stuxnet's Creators Tried to Achieve" (The Langner Group, 2013), `https://archive.org/details/to-kill-a-centrifuge`.

2. Bundesamt für Sicherheit in der Informationstechnik (BSI), "Die Lage der IT-Sicherheit in Deutschland 2014" (The State of IT Security in Germany 2014); see also U.S. Department of Energy (CESER) and Idaho National Laboratory, "Precursor Analysis Report: Cyber Attack on Thyssenkrupp," CyOTE Case Study INL/RPT-22-70679, `https://cyote.inl.gov/content/uploads/24/2025/12/CyOTE-Case-Study_Thyssenkrupp.pdf`.

3. Robert M. Lee, Michael J. Assante, and Tim Conway, "Analysis of the Cyber Attack on the Ukrainian Power Grid: Defense Use Case" (SANS ICS and the Electricity Information Sharing and Analysis Center, March 18, 2016), `https://nsarchive.gwu.edu/sites/default/files/documents/3891751/SANS-and-Electricity-Information-Sharing-and.pdf`. The subsequent 2016 attack used the CRASHOVERRIDE/Industroyer malware; see CISA, "CrashOverride Malware," Alert TA17-163A, `https://www.cisa.gov/news-events/alerts/2017/06/12/crashoverride-malware`.

4. CISA and the Federal Bureau of Investigation (FBI), "DarkSide Ransomware: Best Practices for Preventing Business Disruption from Ransomware Attacks," Joint Cybersecurity Advisory AA21-131A, May 11, 2021, `https://www.cisa.gov/news-events/cybersecurity-advisories/aa21-131a`; U.S. Department of Justice, "Department of Justice Seizes $2.3 Million in Cryptocurrency Paid to the Ransomware Extortionists Darkside," press release, June 7, 2021, `https://www.justice.gov/archives/opa/pr/department-justi`

ce-seizes-23-million-cryptocurrency-paid-ransomware-extortionists-darkside.

5. CISA, "Cyber Incident Reporting for Critical Infrastructure Act of 2022 (CIRCIA)," https://www.cisa.gov/topics/cyber-threats-and-advisories/information-sharing/cyber-incident-reporting-critical-infrastructure-act-2022-circia.

6. Keith Stouffer et al., "SP 800-82 Rev. 3: Guide to Operational Technology (OT) Security" (National Institute of Standards and Technology, September 2023), https://csrc.nist.gov/pubs/sp/800/82/r3/final.

7. CISA, "Cross-Sector Cybersecurity Performance Goals, Version 2.0 (CPG 2.0)," December 2025, https://www.cisa.gov/cybersecurity-performance-goals-2-0-cpg-2-0.

Glossary of Terms

ICS (Industrial Control Systems): A general term that encompasses several types of control systems used in industrial production, including SCADA and DCS.

IT (Information Technology): The use of computers, storage, networking, and other physical devices, infrastructure, and processes to create, process, manage, and store data.

OT (Operational Technology): Hardware and software that detects or causes a change through the direct monitoring and/or control of physical devices, processes, and events in the enterprise. This includes the systems that control and monitor industrial machinery, power grids, water treatment plants, and other critical infrastructure.

SCADA (Supervisory Control and Data Acquisition): A type of ICS used to monitor and control industrial processes in geographically dispersed locations.

DCS (Distributed Control System): A type of ICS used in manufacturing or process control where controllers are distributed throughout the system but supervised by a central computer.

PLC (Programmable Logic Controller): A ruggedized digital computer used for automation of industrial electromechanical processes, such as control of machinery on factory assembly lines, amusement rides, or lighting fixtures.

Cyber-Physical Systems (CPS): Systems where computational algorithms are tightly integrated with physical processes. OT systems are a subset of CPS.

Threat Intelligence: Information about potential or actual threats faced by an organization.

Vulnerability: A weakness in an asset or control that could be exploited by a threat.

Risk: The potential for loss or damage when a threat exploits a vulnerability.

Resilience: The ability of a system to continue to operate, or to quickly recover from disruptions, in the face of cyber incidents.

Air Gap: A security measure that involves physically isolating a network or system from other networks, such as the internet or an untrusted internal network.

Zero Trust: A security model that requires strict identity verification for every person and device trying to access resources on a private network, regardless of whether they are outside or inside the network perimeter.

Supply Chain Attack: A cyberattack that targets less secure elements in the supply chain of a target organization to gain access to the target's systems or data.

Selected Bibliography and Further Reading

1. National Institute of Standards and Technology. (2024). *The NIST Cybersecurity Framework (CSF) 2.0*, NIST CSWP 29 (February 26, 2024). U.S. Department of Commerce.

2. International Society of Automation. (2019). *ISA/IEC 62443 Series of Standards: Security for industrial automation and control systems.*

3. U.S. Cybersecurity and Infrastructure Security Agency. (2022). *Guidance on the Threat Landscape for Operational Technology*. CISA.

4. SANS Institute. (2023). *The SANS 2023 State of OT Security Survey.*

5. Mandiant. (2023). *M-Trends 2023: The 14th Annual Mandiant Incident Response Report.*

6. World Economic Forum. (2023). *Global Cybersecurity Outlook 2023.*

7. Center for Strategic and International Studies. (2022). *The Global Cyber Arms Race.*

8. Brito, Jerry, and Tate Watkins. (2011). *Loving the Cyber Bomb? The Dangers of Threat Inflation in Cybersecurity Policy.* Harvard National Security Journal, 3; Mercatus Center, George Mason University.

9. Lawson, Sean, and Michael K. Middleton. (2019). *Cyber Pearl Harbor: Analogy, Fear, and the Framing of Cyber Security Threats in the United States, 1991–2016.* First Monday, 24(3).

10. Key peer-reviewed venues in the field: *IEEE Transactions on Industrial Informatics*, *Computers & Security*, and the *Journal of Cybersecurity*.

11. Industry threat and market reports drawn on throughout this book: Dragos, *OT/ICS Cybersecurity Year in Review* (annual); Verizon, *Data Breach Investigations Report* (annual); IBM Security, *Cost of a Data Breach Report* (annual); Waterfall Security Solutions, *OT/ICS Cyber Threat Report* (annual); and Gartner, *Forecast: Information Security and Risk Management, Worldwide.*

Notes

THE UNEXAMINED NARRATIVE: OPERATIONAL TECHNOLOGY UNDER SIEGE?

1. Winn Schwartau, who first popularized the phrase "Electronic Pearl Harbor" in 1991 (Computerworld oped and testimony before the U.S. House of Representatives, June 27, 1991); precise attribution of the coinage is debated, https://phibetaiota.net/1991/06/winn-schwartau-digital-pearl-harbor-testimony-to-congress-27-june-1991/
2. Leon Panetta, Secretary of Defense, "cyber Pearl Harbor" address to Business Executives for National Security aboard the USS Intrepid, October 11, 2012, https://archive.defense.gov/transcripts/transcript.aspx?transcriptid=5136
3. Gartner, forecast that worldwide information-security spending will reach roughly $322 billion by 2029 ("Forecast: Information Security, Worldwide," 2025), https://www.gartner.com/en/documents/6998666

THE HISTORICAL RECORD: SCARCITY OF CATASTROPHE

1. Joe Weiss, Applied Control Solutions, "Unfettered" commentary on the undercounting of OT cyber incidents (control-system environments frequently lack the forensic logging needed to distinguish a cyber cause from an equipment fault), https://www.controlglobal.com/blogs/unfettered/blog/55304744/

2. Widely reported attribution of Stuxnet to a joint U.S.–Israeli program reportedly codenamed "Olympic Games.", David E. Sanger, "Obama Order Sped Up Wave of Cyberattacks Against Iran," The New York Times, June 1, 2012, https://www.nytimes.com/2012/06/01/world/middleeast/obama-ordered-wave-of-cyberattacks-against-iran.html

3. Symantec and Langner technical analyses of Stuxnet's Siemens S7 controller / centrifuge-cascade payload, Falliere, O'Murchu & Chien, "W32.Stuxnet Dossier" (Symantec, 2011), https://nsarchive2.gwu.edu/NSAEBB/NSAEBB424/docs/Cyber-044.pdf, and Ralph Langner, "To Kill a Centrifuge" (2013), https://archive.org/details/to-kill-a-centrifuge

4. Attribution of the 2015 and 2016 Ukraine power-grid attacks to the Russian military intelligence group commonly tracked as Sandworm, U.S. Department of Justice, "Six Russian GRU Officers Charged in Connection with Worldwide Deployment of Destructive Malware...," October 19, 2020, https://www.justice.gov/archives/opa/pr/six-russian-gru-officers-charged-connection-worldwide-deployment-destructive-malware-and

5. E-ISAC and SANS Industrial Control Systems analysis of the December 2015 Ukraine grid attack, including the telephonic denial-of-service flooding of the distribution utility's call center, https://nsarchive.gwu.edu/sites/default/files/documents/3891751/SANS-and-Electricity-Information-Sharing-and.pdf (the SANS report analyzes the December 2015 attack; a subsequent December 2016 attack used the CRASHOVERRIDE/Industroyer malware, addressed separately in footnote 10 below).

6. E-ISAC and SANS account of manual restoration of service, in which operators physically drove to the affected substations and closed breakers by hand, https://nsarchive.gwu.edu/sites/default/files/documents/3891751/SANS-and-Electricity-Information-Sharing-and.pdf

7. Dragos and ESET analyses of the December 2016 Industroyer / CrashOverride ICS malware used against a transmission substation north of Kyiv, Dragos, "CRASHOVERRIDE: Analyzing the Malware That Attacks Power Grids" (2017), https://www.dragos.com/resources/whitepaper/crashoverride-analyzing-the-malware-that-attacks-power-grids/;

see also CISA, "CrashOverride Malware," Alert TA17-163A, June 12, 2017, https://www.cisa.gov/news-events/alerts/2017/06/12/crashoverride-malware

8. Mandiant attribution of TRITON/TRISIS to the Russian Central Scientific Research Institute of Chemistry and Mechanics (CNIIHM); U.S. Department of the Treasury sanctions on CNIIHM in connection with the attack, October 2020, https://cloud.google.com/blog/topics/threat-intelligence/triton-attribution-russian-government-owned-lab-most-likely-built-tools

9. FireEye/Mandiant and Schneider Electric analyses of the TRITON intrusion, in which the code inadvertently tripped the Schneider Electric Triconex safety controllers into a protective shutdown, exposing the campaign, https://cloud.google.com/blog/topics/threat-intelligence/triton-actor-ttp-profile-custom-attack-tools-detections

10. President Joseph R. Biden's public statement that there was "no evidence" the Russian government was responsible for the Colonial Pipeline attack, May 2021, https://www.npr.org/2021/06/07/1004050873/u-s-retrieves-some-of-the-colonial-pipeline-ransom

11. Colonial Pipeline and FBI accounts that the DarkSide ransomware affected Colonial's IT and billing systems while the OT pipeline control systems were not compromised, and that the shutdown was a precautionary business decision, https://www.cnbc.com/2021/06/08/colonial-pipeline-ceo-testifies-on-first-hours-of-ransomware-attack.html

12. Reporting that the Colonial Pipeline intrusion used a single legacy VPN account credential with no multi-factor authentication, https://threatpost.com/darkside-pwned-colonial-with-old-vpn-password/166743/

13. U.S. Department of Justice recovery of 63.7 of the approximately 75 bitcoin paid by Colonial Pipeline to the DarkSide ransomware operators, June 2021, https://www.justice.gov/archives/opa/pr/department-justice-seizes-23-million-cryptocurrency-paid-ransomware-extortionists-darkside

14. Pinellas County Sheriff Bob Gualtieri's 2021 press-conference account of the Oldsmar sodium hydroxide setpoint change and the operator's observation of cursor movement on the HMI, `https://edition.cnn.com/2021/02/08/us/oldsmar-florida-hack-water-poison`

15. Oldsmar city manager Al Braithwaite's later (2023) characterization of the incident as a "nonevent," stating the FBI found no evidence of any access from the outside and that it was likely the same employee "actually banging on his keyboard.", `https://www.wusf.org/local-state/2023-04-13/official-hack-oldsmar-city-water-treatment-plant-2021`

16. CyberScoop, reporting that "the FBI was not able to confirm that this incident was initiated by a targeted cyber intrusion.", `https://cyberscoop.com/water-oldsmar-incident-cyberattack/`

17. Sheriff Bob Gualtieri's statements that the altered sodium-hydroxide setpoint "would have taken from 24 to 36 hours to take effect" and would have triggered the plant's pH alarms before any tainted water reached a customer, `https://edition.cnn.com/2021/02/08/us/oldsmar-florida-hack-water-poison`

18. CISA reporting on the Cyber Av3ngers compromise of internet-exposed Unitronics programmable logic controllers using the default password "1111," in which the intruders changed the HMI screens to display a political message, 2023–2024, `https://thehackernews.com/2023/11/iranian-hackers-exploit-plcs-in-attack.html`

19. German Federal Office for Information Security (BSI), Die Lage der IT-Sicherheit in Deutschland 2014 (2014 annual report), approximately p. 31 — the anonymous, undated account of damage to a German steel mill blast furnace, `https://www.securityweek.com/cyberattack-german-steel-plant-causes-significant-damage-report/`

20. Idaho National Laboratory, 2007 Aurora Generator Test, a controlled experiment in which a cyber technique was used to open and reclose a diesel generator's breakers out of phase, destroying the machine under controlled conditions, U.S. Department of Homeland Security / Idaho National Laboratory, Aurora Generator Test (March 2007); DHS Control Systems Security Program records

released under FOIA (2014), as reported by MuckRock, https://www.muckrock.com/news/archives/2016/nov/14/aurora-generator-test-homeland-security/

21. CISA et al., joint cybersecurity advisory AA24-038A, on PRC state-sponsored actor Volt Typhoon pre-positioning on the IT networks of U.S. critical infrastructure organizations, February 2024, https://www.cisa.gov/news-events/cybersecurity-advisories/aa24-038a

22. U.S. Department of Justice, indictment of five People's Liberation Army (Unit 61398) officers for cyber-espionage against Westinghouse, U.S. Steel, Alcoa, and other firms, May 19, 2014, https://www.justice.gov/opa/pr/us-charges-five-chinese-military-hackers-cyber-espionage-against-us-corporations-and-labor; see also CISA Advisory AA22-083A on Russian state-sponsored intrusions into the U.S. energy sector, https://www.cisa.gov/news-events/cybersecurity-advisories/aa22-083a

23. JBS and Reuters/Toyota disclosures of ransomware-driven operational shutdowns in 2021–2022, including Toyota's suspension of its 14 domestic plants after a ransomware attack on its supplier Kojima Industries (Reuters, Feb. 28, 2022; Toyota corporate statement).

24. Idaho National Laboratory, 2007 Aurora Generator Test, a controlled laboratory experiment whose footage has been replayed as if it documented a real attack on the power grid, U.S. Department of Homeland Security / Idaho National Laboratory, Aurora Generator Test (March 2007); DHS Control Systems Security Program records released under FOIA (2014), as reported by MuckRock, https://www.muckrock.com/news/archives/2016/nov/14/aurora-generator-test-homeland-security/

25. CISA et al., joint cybersecurity advisory AA24-038A, on Volt Typhoon pre-positioning in U.S. critical infrastructure networks, February 2024, https://www.cisa.gov/news-events/cybersecurity-advisories/aa24-038a

26. Sheriff Bob Gualtieri's statements on the 24-to-36-hour delay before the altered Oldsmar setpoint would take effect and on the pH alarms positioned to fire before any harm, `https://edition.cnn.com/2021/02/08/us/oldsmar-florida-hack-water-poison`

27. Oldsmar city manager Al Braithwaite's "nonevent" characterization; CyberScoop reporting on the FBI's inability to confirm the incident was initiated by a targeted cyber intrusion, `https://cyberscoop.com/water-oldsmar-incident-cyberattack/`

28. CISA reporting on the Cyber Av3ngers compromise of internet-exposed Unitronics PLCs reachable with the default password "1111" and the resulting HMI defacement, 2023–2024, `https://thehackernews.com/2023/11/iranian-hackers-exploit-plcs-in-attack.html`

29. German Federal Office for Information Security (BSI), 2014 annual report — the anonymous steel mill paragraph, naming no company, citing no malware, and quantifying no damage, `https://www.securityweek.com/cyberattack-german-steel-plant-causes-significant-damage-report/`

30. Colonial Pipeline and FBI accounts that the OT pipeline controls were not compromised and that the line was taken offline as a precautionary, billing-related business decision, `https://www.cnbc.com/2021/06/08/colonial-pipeline-ceo-testifies-on-first-hours-of-ransomware-attack.html`

31. CISA et al., joint cybersecurity advisory AA24-038A, on Volt Typhoon, February 2024, `https://www.cisa.gov/news-events/cybersecurity-advisories/aa24-038a`

32. Sheriff Bob Gualtieri's delay-and-alarm statements; Oldsmar city manager Al Braithwaite's "nonevent" characterization; CyberScoop reporting on the unconfirmed nature of the intrusion, `https://cyberscoop.com/water-oldsmar-incident-cyberattack/`

33. FireEye/Mandiant and Schneider Electric analyses of the TRITON intrusion, exposed because the targeted Triconex safety system tripped the plant to a safe state, `https://cloud.google.com/blog/topics/threat-intelligence/triton-actor-ttp-profile-custom-attack-tools-detections`

34. Oldsmar city manager Al Braithwaite's internal-explanation characterization of the incident as likely "the same employee... banging on his keyboard" rather than a foreign adversary (2023), `https://www.route-fifty.com/cybersecurity/2023/03/florida-city-water-cyber-incident-allegedly-caused-employee-error/384267/`

35. Joe Weiss, Applied Control Solutions, "Unfettered" commentary arguing that OT physical-impact incidents are under-counted because control-system environments lack the forensic instrumentation to distinguish a malicious manipulation from an equipment fault or a human error, `https://www.controlglobal.com/blogs/unfettered/blog/55304744/`

36. CISA reporting on the Unitronics default-password ("1111") exposure exploited by Cyber Av3ngers; reporting that the Colonial Pipeline intrusion used a single VPN credential with no multi-factor authentication, `https://thehackernews.com/2023/11/iranian-hackers-exploit-plcs-in-attack.html`

37. The multi-year gap between the initial 2021 Oldsmar attack narrative (Sheriff Gualtieri's press conference) and the later "nonevent" walk-back by Oldsmar city manager Al Braithwaite (2023), `https://www.wusf.org/local-state/2023-04-13/official-hack-oldsmar-city-water-treatment-plant-2021`

38. U.S. Department of the Treasury sanctions on the Russian institute CNIIHM over TRITON, October 2020; Mandiant's technical attribution assessment, `https://home.treasury.gov/news/press-releases/sm1162`

39. President Joseph R. Biden's "no evidence" statement regarding Russian government responsibility for the Colonial Pipeline attack, May 2021, `https://www.npr.org/2021/06/07/1004050873/u-s-retrieves-some-of-the-colonial-pipeline-ransom`

40. The Oldsmar attribution reversal — from Sheriff Gualtieri's confident 2021 public assertion of an outside intrusion to the unconfirmed "nonevent" characterization roughly two years later, `https://www.wusf.org/local-state/2023-04-13/official-hack-oldsmar-city-water-treatment-plant-2021`

41. Analysis of DarkSide's ransomware-as-a-service structure, which separates the malware's developers from the affiliates who operate it, complicating attribution of the Colonial Pipeline attack, `https://www.eff.org/deeplinks/2021/05/faq-darkside-ransomware-group-and-colonial-pipeline`

THE MAZE OF U.S. GOVERNANCE: OVERLAPPING MANDATES AND PARALYSIS

1. National Institute of Standards and Technology, *Guide to Operational Technology (OT) Security*, NIST Special Publication 800-82 Revision 3 (September 2023), `https://csrc.nist.gov/pubs/sp/800/82/r3/final`

2. The Department of Energy serves as the Sector Risk Management Agency for the Energy Sector. The role originates in Presidential Policy Directive 21 (2013), which designated "Sector-Specific Agencies"; the National Defense Authorization Act for FY2021 §9002 (Pub. L. 116-283) recodified these as Sector Risk Management Agencies (6 U.S.C. §665d). The Transportation Security Administration's pipeline-security authority derives from the Aviation and Transportation Security Act of 2001 (49 U.S.C. §114). `https://www.law.cornell.edu/uscode/text/6/665d`

3. U.S. Environmental Protection Agency, data on community water systems (roughly 50,000 nationwide), Safe Drinking Water Information System, `https://www.epa.gov/groundwater-and-drinking-water/information-about-public-water-systems`

4. Cybersecurity and Infrastructure Security Agency, mission and constituency of critical-infrastructure owners and operators, `https://www.cisa.gov/about`

5. CISA's annual appropriations grew from roughly $2 billion at the agency's standup (FY2019–FY2020) to roughly $3 billion by FY2025 (FY2024 enacted approximately $2.87 billion). See CISA Congressional Budget Justifications and Congressional Research Service, "Comparing DHS Component Funding" (R48115), `https://www.congress.gov/crs-product/R48115`

6. North American Electric Reliability Corporation, Critical Infrastructure Protection (CIP) Reliability Standards, mandatory and enforceable under the oversight of the Federal Energy Regulatory Commission per §215 of the Federal Power Act (16 U.S.C. §824o), `https://www.ferc.gov/industries-data/electric/industry-activities/reliability-standards`

7. U.S. Environmental Protection Agency, interpretive memorandum "Addressing Public Water System Cybersecurity in Sanitary Surveys" (March 3, 2023), withdrawn October 11, 2023 following litigation, `https://www.epa.gov/waterresilience/epa-cybersecurity-water-sector`

8. Transportation Security Administration, Security Directive Pipeline-2021-01 (May 28, 2021) and Pipeline-2021-02 (July 2021), issued following the Colonial Pipeline ransomware incident, `https://www.federalregister.gov/documents/2021/09/24/2021-20738/ratification-of-security-directive`

9. Transportation Security Administration's subsequent shift toward a performance-based (rather than prescriptive) model in Security Directive Pipeline-2021-02C (July 2022), `https://www.tsa.gov/news/press/releases/2022/07/21/tsa-revises-and-reissues-cybersecurity-requirements-pipeline-owners`

10. Cyber Incident Reporting for Critical Infrastructure Act of 2022 (CIRCIA), enacted as Division Y of the Consolidated Appropriations Act, 2022 (Pub. L. 117-103); CISA implementing rulemaking ongoing, `https://www.cisa.gov/topics/cyber-threats-and-advisories/information-sharing/cyber-incident-reporting-critical-infrastructure-act-2022-circia`

THE VENDOR ECOSYSTEM: PROFITING FROM FEAR

1. Ian Levy, then Technical Director, UK National Cyber Security Centre (GCHQ), remarks reported Feb. 2017, `https://www.infosecurity-magazine.com/news/cyber-expert-vendor-fud-distorts/`

2. Bruce Schneier, *Beyond Fear: Thinking Sensibly About Security in an Uncertain World* (Copernicus Books, 2003); coinage of "security theater.", https://www.schneier.com/tag/security-theater/

3. Bruce Schneier, *Beyond Fear: Thinking Sensibly About Security in an Uncertain World* (Copernicus Books, 2003), on the human tendency to overestimate rare, spectacular risks and underestimate commonplace ones, https://www.schneier.com/tag/security-theater/

4. Gen. Keith Alexander, then Director of the National Security Agency, remarks at the American Enterprise Institute, July 2012 ("the greatest transfer of wealth in history"), https://foreignpolicy.com/2012/07/09/nsa-chief-cybercrime-constitutes-the-greatest-transfer-of-wealth-in-history/

5. IronNet Cybersecurity, founded 2014; reported consulting fees of up to $1 million per month (reported 2014), https://www.ibtimes.com/former-nsa-director-keith-alexander-denies-revealing-us-intelligence-secrets-massive-1650662

6. IronNet public listing via SPAC merger at a valuation of roughly $1.2 billion, Aug. 2021, IronNet, Inc., Form 8-K (U.S. Securities and Exchange Commission, EDGAR), August 2021, https://www.sec.gov/Archives/edgar/data/0001777946/000119312521263381/d216367dex991.htm

7. IronNet share price peaked at $31.26 intraday on September 17, 2021, https://stockanalysis.com/stocks/irntq/

8. IronNet downgraded its annual recurring revenue (ARR) projections by roughly 60 percent in December 2021, https://www.securityweek.com/collapse-of-national-security-elites-cyber-firm-leaves-bitter-wake/

9. IronNet shareholder lawsuit allegations regarding a roughly $10 million U.S. Navy-contractor arrangement and a more than $22 million Kuwait deal that, per the complaint, had "no realistic chance of closing in FY22," https://www.bankinfosecurity.com/ironnet-eyes-bankruptcy-as-ex-worker-backs-class-action-suit-a-20729

10. IronNet shareholder securities-fraud class action settled for approximately $6.6 million ($6.625 million), `https://www.bankinfosecurity.com/ironnet-settles-securities-fraud-class-action-suit-for-66m-a-26473`

11. IronNet stock fell to $0.26, was delisted from the NYSE in 2023, and the company filed for Chapter 11 bankruptcy in October 2023, In re IronNet, Inc., No. 23-11710 (Bankr. D. Del., Oct. 12, 2023); see also IronNet, Inc., Form 8-K, U.S. Securities and Exchange Commission, `https://www.sec.gov/Archives/edgar/data/0001777946/000119312523255052/d549915d8k.htm`

12. Bloomberg, "Former NSA Chief Keith Alexander Caught SPAC Fever, and Investors Got Burned.", `https://www.bloomberg.com/news/articles/2024-01-18/former-nsa-chief-keith-alexander-caught-spac-fever-and-investors-got-burned`

13. Industrial-equipment makers as investors in OT-security firms: Schneider Electric has backed both Claroty and Nozomi Networks; Honeywell and Mitsubishi Electric have backed Nozomi; Rockwell Automation and Siemens (Next47) have backed Claroty. Claroty Series E (Rockwell Automation co-lead), `https://claroty.com/press-releases/claroty-announces-rockwell-automation-as-co-leader-of-400-million-series-e-funding`

14. Mitsubishi Electric acquisition of Nozomi Networks, reported at roughly $1 billion, 2025, `https://www.cybersecuritydive.com/news/mitsubishi-electric-buy-nozomi-networks-1b/759637/`

15. Rockwell Automation acquisition of the OT-security firm Verve Industrial, 2023, `https://www.businesswire.com/news/home/20231023944076/en/Rockwell-Automation-Signs-Agreement-to-Acquire-Verve-Industrial-Protection`

16. Honeywell acquisition of the OT-security vendor SCADAfence, 2023, `https://www.prnewswire.com/news-releases/honeywell-to-acquire-scadafence-strengthening-its-cybersecurity-software-portfolio-301872646.html`

17. The venture arms of Koch Industries (Koch Disruptive Technologies) and Saudi Aramco (Saudi Aramco Energy Ventures) among the backers of Dragos in its

Series C round, December 2020, https://www.businesswire.com/news/home/20201208005280/en/

18. U.S. Securities and Exchange Commission cybersecurity incident-disclosure rule requiring disclosure within four business days of a materiality determination, adopted 2023, https://www.sec.gov/newsroom/press-releases/2023-139

19. Directive (EU) 2022/2555 (NIS2 Directive), 14 December 2022, expanding cybersecurity risk-management and incident-reporting obligations across "essential" and "important" entities, https://eur-lex.europa.eu/eli/dir/2022/2555/oj

20. Jerry Brito and Tate Watkins, "Loving the Cyber Bomb? The Dangers of Threat Inflation in Cybersecurity Policy," Mercatus Center, George Mason University (2011), later issued through the *Harvard National Security Journal*, https://www.mercatus.org/research/working-papers/loving-cyber-bomb-dangers-threat-inflation-cybersecurity-policy

21. Jerry Brito and Tate Watkins, "Loving the Cyber Bomb? The Dangers of Threat Inflation in Cybersecurity Policy," Mercatus Center, George Mason University (2011): a cyber-industrial complex "may serve to not only supply cybersecurity solutions to the federal government, but to drum up demand for them as well," https://www.mercatus.org/research/working-papers/loving-cyber-bomb-dangers-threat-inflation-cybersecurity-policy

22. Jerry Brito and Tate Watkins, "The Cybersecurity-Industrial Complex," Reason, July 25, 2011: "Washington is filled with people who have a vested interest in conflating and inflating the threats to our digital security," https://reason.com/2011/07/25/the-cybersecurity-industrial-c/

23. Thoma Bravo acquired Imperva for approximately $2.1 billion (2019) and sold it to Thales for approximately $3.6 billion (2023), https://www.thomabravo.com/press-releases/thoma-bravo-completes-sale-of-imperva-to-thales

24. Insight Partners and Armis: a stake at a valuation of roughly $1.1 billion (2020), marked to roughly $3.4 billion (2021), with an agreed sale to ServiceNow at $7.75

billion (Dec. 2025), https://www.insightpartners.com/ideas/insight-partners-acquires-armis-at-a-1-1-billion-valuation/, https://newsroom.servicenow.com/press-releases/details/2026/ServiceNow-completes-Armis-acquisition-closing-the-gap-between-asset-visibility-and-cyber-risk/default.aspx

25. Thoma Bravo managing approximately $181 billion in assets under management (as of June 30, 2025), Thoma Bravo, firm assets-under-management disclosure, https://www.thomabravo.com

26. Thoma Bravo acquired Proofpoint for $12.3 billion (2021), https://www.globenewswire.com/news-release/2021/04/26/2216810/0/en/Proofpoint-Enters-Definitive-Agreement-to-be-Acquired-by-Thoma-Bravo-in-12-3-Billion-Transaction.html

27. Thoma Bravo 2022 identity-security acquisitions: SailPoint for $6.9 billion, Ping Identity for $2.8 billion, and ForgeRock for $2.3 billion (subsequently merged into Ping), Thoma Bravo, "ForgeRock to be Acquired by Thoma Bravo for $2.3B," October 11, 2022, https://www.thomabravo.com/press-releases/forgerock-to-be-acquired-by-thoma-bravo-for-2.3b

28. Thoma Bravo took Sophos private for $3.9 billion (2020) and agreed to acquire Darktrace for $5.3 billion (2024), https://cyberscoop.com/sophos-thoma-bravo-acquisition/, https://www.prnewswire.com/news-releases/darktrace-announces-formal-completion-of-its-acquisition-by-thoma-bravo-302263852.html

29. Venture-capital investment into cybersecurity: roughly $22.8 billion (2021), $15.3 billion (2022), $8.1 billion (2023), and $11.6 billion (2024), https://news.crunchbase.com/cybersecurity/big-rounds-comeback-4q-eoy-2024/

30. Gartner, forecast that worldwide information-security spending will reach roughly $322 billion by 2029 ("Forecast: Information Security, Worldwide," 2025), https://www.gartner.com/en/documents/6998666

31. McKinsey & Company, estimate of a $1.5 trillion to $2 trillion addressable cybersecurity market at roughly 10 percent current penetration (October 2022),

https://www.mckinsey.com/capabilities/risk-and-resilience/our-insights/cybersecurity/new-survey-reveals-2-trillion-dollar-market-opportunity-for-cybersecurity-technology-and-service-providers

POLICY FRAMEWORKS BUILT ON MYTH: A CRITICAL EXAMINATION

1. Executive Order 13636, "Improving Critical Infrastructure Cybersecurity," February 12, 2013 (78 FR 11739), https://www.federalregister.gov/documents/2013/02/19/2013-03915/improving-critical-infrastructure-cybersecurity
2. Leon Panetta, "Remarks on Cybersecurity to the Business Executives for National Security," New York, Oct. 11, 2012, U.S. Department of Defense transcript; archived at National Security Archive, https://nsarchive.gwu.edu/document/21479-document-78
3. Adm. Michael S. Rogers, testimony before the House (Select) Permanent Intelligence Committee, November 20, 2014 ("it is only a matter of when, not if, we are going to see something dramatic"); the characterization of the prediction's horizon as "before 2025" is the author's reading of the testimony's implied timeframe, https://edition.cnn.com/2014/11/20/politics/nsa-china-power-grid
4. Jen Easterly, testimony before the House Select Committee on the Chinese Communist Party, Jan. 31, 2024; CISA; reported by Defense One, https://www.defenseone.com/technology/2024/02/chinese-hacking-operations-have-entered-far-more-dangerous-phase-us-warns/393843/
5. James R. Clapper, Director of National Intelligence, "Statement for the Record: Worldwide Cyber Threats," House Permanent Select Committee on Intelligence, September 10, 2015, rejecting a "Cyber Armageddon" scenario in favor of "an ongoing series of low-to-moderate level cyber attacks ... which will impose cumulative costs," https://docs.house.gov/meetings/IG/IG00/20150910/103797/HHRG-114-IG00-Wstate-ClapperJ-20150910.PDF

6. Winn Schwartau first popularized the "electronic Pearl Harbor" framing in testimony before the House Subcommittee on Technology and Competitiveness, June 27, 1991, and a 1991 Computerworld oped ("Fighting Terminal Terrorism"); precise attribution of the coinage is debated, https://phibetaiota.net/1991/06/winn-schwartau-digital-pearl-harbor-testimony-to-congress-27-june-1991/

7. Richard A. Clarke, quoted in Marjorie Censer, "Little cyber attacks have big effects," Washington Post, May 7, 2012 ("the death of a thousand cuts in the little Pearl Harbors that are happening every day"); cited in Lawson & Middleton (2016), https://www.seanlawson.net/wp-content/uploads/2017/02/LawsonMiddleton-CyberPearlHarborEssay.pdf

8. Sean Lawson and Michael K. Middleton, "Cyber Pearl Harbor: Analogy, Fear, and the Framing of Cyber Security Threats in the United States, 1991–2016," University of Utah, 2016, https://www.seanlawson.net/wp-content/uploads/2017/02/LawsonMiddleton-CyberPearlHarborEssay.pdf

9. Executive Order 13636, "Improving Critical Infrastructure Cybersecurity," February 12, 2013 (78 FR 11739), https://www.federalregister.gov/documents/2013/02/19/2013-03915/improving-critical-infrastructure-cybersecurity

10. James R. Clapper, Director of National Intelligence, "Statement for the Record: Worldwide Cyber Threats," House Permanent Select Committee on Intelligence, September 10, 2015, https://docs.house.gov/meetings/IG/IG00/20150910/103797/HHRG-114-IG00-Wstate-ClapperJ-20150910.PDF

THE MYTH OF IMMINENT CATASTROPHE: A STATISTICAL REALITY CHECK

1. Dragos, "OT/ICS Cybersecurity Year in Review 2024" (8th Annual), Feb. 2025, https://www.dragos.com/blog/dragos-8th-annual-ot-cybersecurity-year-in-review-is-now-available

2. Dragos, "OT/ICS Cybersecurity Year in Review 2024" (8th Annual), Feb. 2025 (threat group BAUXITE; FrostyGoop and Fuxnet malware families), https:

//www.dragos.com/blog/dragos-8th-annual-ot-cybersecurity-year-in-review-is-now-available

3. Robert M. Lee (CEO, Dragos), quoted in Cybersecurity Dive, January 2023, https://www.cybersecuritydive.com/news/industrial-ics-vulnerabilities-dragos/641418/

4. IBM, "X-Force Threat Intelligence Index 2024" (manufacturing the most-attacked industry, ~26% of incidents), https://www.ibm.com/think/x-force/2024-x-force-threat-intelligence-index

5. WIRED, reporting on Dragos analysis (approximately three instances of in-the-wild, physically destructive ICS malware: Stuxnet, Industroyer/CRASHOVERRIDE Triton), https://www.wired.com/story/russia-ukraine-cyberattack-power-grid-blackout-destruction/

6. WIRED (Andy Greenberg) and Dragos document only a handful of in-the-wild cyberattacks built for destructive, disruptive, or safety-relevant cyber-physical effects in the fifteen years since Stuxnet — principally Stuxnet (2010), the two Ukrainian grid attacks (2015 and 2016), and Triton (2017); the exact count varies with inclusion criteria, https://www.wired.com/story/russia-ukraine-cyberattack-power-grid-blackout-destruction/

7. PipeDream/Incontroller, a modular ICS attack framework publicly disclosed in April 2022 (CISA Alert AA22-103A; Dragos threat group CHERNOVITE) — a capability discovered before deployment and never used in the wild, https://www.cisa.gov/news-events/cybersecurity-advisories/aa22-103a

8. Waterfall Security Solutions / ICSSTRIVE, "Threat Report" (57 incidents with physical consequences worldwide in 2025, down ~25% year over year), https://waterfall-security.com/ot-insights-center/ot-cybersecurity-insights-center/2025-threat-report-ot-cyberattacks-with-physical-consequences/

9. Dale Peterson, decomposition of the Waterfall Security / ICSSTRIVE data (nation-state OT-physical attacks worldwide = 3 in 2025, down from 6 in 2024), dale-peterson.com, https://dale-peterson.com/2026/03/31/numbe

r-of-cyber-incidents-with-a-physical-impact-down-in-2025/

10. Verizon, "2024 Data Breach Investigations Report" (DBIR) (30,458 security incidents; 10,626 confirmed data breaches), https://www.verizon.com/business/resources/reports/dbir/
11. Identity Theft Resource Center (ITRC) (approximately 60 billion exposed records tracked since 2005), https://www.idtheftcenter.org/publication/2024-data-breach-report/
12. IBM, "Cost of a Data Breach Report 2024" (global average breach cost $4.88 million), https://www.ibm.com/think/insights/whats-new-2024-cost-of-a-data-breach-report
13. Joe Weiss, "Unfettered" blog, ControlGlobal (OT-incident undercounting argument), https://www.controlglobal.com/blogs/unfettered/blog/55304744/
14. Maroochy Shire, Queensland, Australia (2000): Vitek Boden, a disgruntled former contractor, used stolen radio and SCADA equipment to release sewage — an insider attack, not a remote network intrusion, https://www.mitre.org/sites/default/files/pdf/08_1145.pdf
15. Diesel emissions "defeat devices" (e.g., the Volkswagen case): manufacturer fraud manipulating engine-control software, not an external attacker compromising a victim's infrastructure, https://www.epa.gov/vw/learn-about-volkswagen-violations

INTERNATIONAL PERSPECTIVES: DIVERSE APPROACHES TO OT SECURITY

1. Ian Levy, then Technical Director, UK National Cyber Security Centre, remarks comparing parts of the threat-intelligence industry to "medieval witchcraft" and a "magic amulet," February 2017, https://www.infosecurity-magazine.com/news/cyber-expert-vendor-fud-distorts/

THE ILLUSION OF CONTROL: SYSTEMIC VULNERABILITIES UNMASKED

1. Robert M. Lee, CEO of Dragos, statement that "there have been zero known ICS vulnerabilities ever leveraged in any ICS cyberattack," Cybersecurity Dive, January 2023, `https://www.cybersecuritydive.com/news/industrial-ics-vulnerabilities-dragos/641418/`

2. Robert M. Lee / Dragos, quoted in "Industrial organizations may worry too much about ICS vulnerabilities," Cybersecurity Dive, January 2023, `https://www.cybersecuritydive.com/news/industrial-ics-vulnerabilities-dragos/641418/`

3. The 2017 Triton/Trisis attack on a petrochemical plant's Safety Instrumented System, discovered only when it accidentally tripped the plant into a safe shutdown; attributed to a state-backed actor (CNIIHM, later sanctioned by the U.S. Treasury), `https://cloud.google.com/blog/topics/threat-intelligence/triton-attribution-russian-government-owned-lab-most-likely-built-tools`

4. The Cyber Av3ngers 2023–24 compromise of internet-exposed programmable logic controllers using the default password "1111," resulting in screen defacement with no loss of service, `https://thehackernews.com/2023/11/iranian-hackers-exploit-plcs-in-attack.html`

5. Colonial Pipeline, 2021: initial access via a single legacy VPN credential with no multi-factor authentication, `https://threatpost.com/darkside-pwned-colonial-with-old-vpn-password/166743/`

6. The Colonial Pipeline ransomware incident of May 2021 was attributed to a legacy VPN account not in active use, accessed via a password later found in a batch of leaked credentials on the dark web, though a definitive link to a specific prior breach was not publicly established; the ransomware affected business IT systems, and Colonial shut down OT operations as a precautionary measure.

REDUCING SYSTEMIC VULNERABILITIES THROUGH GOVERNANCE REFORM

1. U.S. Securities and Exchange Commission, 2023 rule requiring public companies to disclose material cybersecurity incidents within four business days of determining materiality, `https://www.sec.gov/newsroom/press-releases/2023-139`

2. Directive (EU) 2022/2555 (NIS2 Directive), expanding cybersecurity and incident-reporting obligations across critical and important sectors, `https://eur-lex.europa.eu/eli/dir/2022/2555/oj`

3. U.S. Securities and Exchange Commission, 2023 four-day cybersecurity incident-disclosure rule, `https://www.sec.gov/newsroom/press-releases/2023-139`

4. Directive (EU) 2022/2555 (NIS2 Directive), incident-reporting duties across European critical and important sectors, `https://eur-lex.europa.eu/eli/dir/2022/2555/oj`

5. U.S. Securities and Exchange Commission, 2023 four-day breach-disclosure rule, `https://www.sec.gov/newsroom/press-releases/2023-139`; and Directive (EU) 2022/2555 (NIS2 Directive), `https://eur-lex.europa.eu/eli/dir/2022/2555/oj`

BEYOND FEAR: SHIFTING FROM PERCEIVED THREATS TO SYSTEMIC RESILIENCE

1. Robert M. Lee, CEO of Dragos, statement that there have been zero known ICS vulnerabilities ever leveraged in any ICS cyberattack, Cybersecurity Dive, January 2023, `https://www.cybersecuritydive.com/news/industrial-ics-vulnerabilities-dragos/641418/`

2. WIRED/Dragos compilations indicate only a handful of in-the-wild attacks built for destructive, disruptive, or safety-relevant cyber-physical effects in the fifteen years since Stuxnet — chiefly Stuxnet, the two Ukrainian grid attacks,

and Triton, `https://www.wired.com/story/russia-ukraine-cyberattack-power-grid-blackout-destruction/`

3. Dale Peterson, Waterfall Security / ICSSTRIVE 2025 threat-report decomposition counting OT incidents with physical consequences, finding worldwide nation-state OT physical attacks in the single digits and declining year over year, `https://dale-peterson.com/2026/03/31/number-of-cyber-incidents-with-a-physical-impact-down-in-2025/`

CONCLUSION: SECURING AMERICA'S FUTURE IN A COMPLEX WORLD

1. Jerry Brito and Tate Watkins, "Loving the Cyber Bomb? The Dangers of Threat Inflation in Cybersecurity Policy," Mercatus Center, George Mason University / Harvard National Security Journal, 2011, `https://www.mercatus.org/research/working-papers/loving-cyber-bomb-dangers-threat-inflation-cybersecurity-policy`

2. The security firm Armis, whose valuation rose from roughly $1.1 billion (Insight Partners, 2020) to its $7.75 billion acquisition by ServiceNow, completed April 20, 2026, over about five years; developed in full in Chapter 4, `https://www.insightpartners.com/ideas/insight-partners-acquires-armis-at-a-1-1-billion-valuation/`, `https://newsroom.servicenow.com/press-releases/details/2026/ServiceNow-completes-Armis-acquisition-closing-the-gap-between-asset-visibility-and-cyber-risk/default.aspx`

3. Fifteen years after Stuxnet, the independently verified list of cyberattacks that produced confirmed cyber-physical effects remains small and category-dependent: Stuxnet caused physical equipment destruction; the 2015 and 2016 Ukraine grid attacks caused bounded service disruptions; and TRITON targeted a safety-instrumented system, discovered after a fail-safe trip. The exact membership varies with inclusion criteria — some lists substitute the disputed 2014 German steel-mill report, and bounding cases such as FrostyGoop (a limited ICS-driven heating outage) and PIPEDREAM/Incontroller (demonstrated capability without public deployment) sharpen rather than overturn the picture,

https://www.wired.com/story/russia-ukraine-cyberattack-power-grid-blackout-destruction/

4. Information-technology breaches run to tens of thousands of incidents annually (e.g., Verizon 2026 Data Breach Investigations Report, which analyzed more than 31,000 security incidents in a single year), https://www.verizon.com/business/resources/reports/dbir/

5. See Chapter 13 (the consequence-centric resilience paradigm).

6. See Chapter 12 (the independent-oversight model).

About the Author

Dr. Daniel Ward, Ph.D., is an Assistant Professor of Information Technology at Southern New Hampshire University. A multidisciplinary scholar, his research interests include both Operational Technology cybersecurity and Cyberpsychology. With over 15 years of leadership experience in the manufacturing technology sector, Dr. Ward brings a practical, "systems thinking" approach to his writing. This technical foundation is uniquely balanced by his background in the behavioral sciences, including advanced degrees in Psychology and Cybersecurity. He is the author of works exploring the complex intersection of technology and society, ranging from the impact of Artificial Intelligence on political polarization to the intimate dynamics of maintaining healthy relationships while navigating bipolar disorder. Dr. Ward holds a Ph.D. in Information Technology, an MS in Psychology, an MS in Cybersecurity and Information Assurance, an MBA, and a Bachelor of Science in Behavioral Science, using his diverse expertise to diagnose societal and policy fractures and offer pathways toward resilience.

www.ingramcontent.com/pod-product-compliance
Lightning Source LLC
Chambersburg PA
CBHW031336230926
51848CB00002B/253
* 9 7 8 1 9 7 2 8 4 2 0 1 0 *